River
Rescue

River
Rescue

Les Bechdel and
Slim Ray

Photographs by Slim Ray
Sketches by Jan AtLee

APPALACHIAN MOUNTAIN CLUB
BOSTON, MASSACHUSETTS

RIVER RESCUE
by Les Bechdel and Slim Ray
Copyright © 1985

Editorial direction: Aubrey Botsford
Production: Nancy Maynes
Book design: Joyce C. Weston
Cover photograph: Slim Ray

10 9 8 7 6 5   88

ISBN: 910146-55-1

Contents

FOREWORD   1

PROLOGUE: TWO RESCUES   3

INTRODUCTION   7

Chapter 1: River Sense   9
Characteristics of Whitewater Rivers   11
Hazards   12
Rating the Rapids   22
Preparation   24
Leadership   30
Saying No   30
The "What If?" Factor   31
Conclusion   32

Chapter 2: Equipment   33
Personal Safety Equipment   34
Clothing   42
Throw Ropes and Throw Bags   45
Boats and Rafts   47
Conclusion   55

Chapter 3: Self-Rescue   57
Strainers   63
Entrapment   63

Holes 64
Conclusion 67

Chapter 4: Rescue by Rope 69
The Throwing Rescue 69
Throwing Techniques 75
Multiple Swimmers 80
Tag-Line Rescues 80
Strong-Swimmer Rescues 88
Conclusion 92

Chapter 5: Boat-Based Rescue 93
The Eskimo Rescue 93
Equipment Retrieval 94
Rescuing Swimmers 99
The Telfer Lower 104
Conclusion 112

Chapter 6: Entrapments and Extrications 113
Entrapments and Boat Pins 114
Entrapment Rescues 116
Recovery of Pinned Boats 119
Rigging 120
The Force of the Current 125
Haul Systems 127
Conclusion 135

Chapter 7: Vertical Rescue 139
Bridge Lowers 139
Tyroleans 146
Helicopter Rescue 153
Helicopter Evacuation 158
Conclusion 160

Chapter 8: Organization for Rescue 164
The Rescue Process 164
Leadership 166
Methods of Rescue 167
Team Organization 168

If the First Attempt Fails   170
Thinking the Unthinkable — The Failed Rescue   171
Liability   173
Reactions   173

Chapter 9: Patient Care and Evacuation   176
The Initial Contact   177
Cardiopulmonary Resuscitation (CPR)   178
Hypothermia   180
Shoulder Dislocations   182
Evacuation Techniques   183
Moving the Litter   189
Conclusion   194

Chapter 10: The Professionals   195
The Rescue Professional   196
The River Professional   202
Conclusion   206

AFTERWORD   207

APPENDICES   209
Appendix A: International Scale of River Difficulty   209
Appendix B: Universal River Signals   210
Appendix C: The Force of the Water   211
Appendix D: Useful Knots   212
Appendix E: Cold Water Survival Chart   213
Appendix F: First Aid Kit Contents   214
Appendix G: Ground-to-Air Signals for Survivors   215
Appendix H: Cardiopulmonary Resuscitation (CPR)   218
Appendix I: Symptoms of Hypothermia   220

INDEX   221

ABOUT THE AMC   227

FOREWORD
by
Charlie Walbridge
ACA Safety Chairman

In the past twenty years whitewater sport has grown tremendously. With that growth has come a vast improvement in both equipment and skills. High-quality factory-made products have largely replaced the home-made gear with which many of us began. Well-organized clinics run by clubs and commercial outfitters have largely replaced the informal "school of hard broaches and long swims," with today's beginners learning more in a month than we did in a year. It is now possible for gutsy newcomers to attempt runs in their first year that frustrated top experts in the 1960s and 1970s. This has led to the opening up of new rivers of incredible difficulty and the performance of wild "hot dog" maneuvers not even thought of as little as ten years ago. The skill level of the average paddler today is higher than ever before in all areas but one: the ability to make effective rescues.

The reasons for this are simple. Before the 1970s, when fiberglass boats were not as prevalent as they later became, one's career in whitewater paddling used to begin with a long apprenticeship in open canoes. The rivers may have been easy by today's standards, but an aluminum canoe with no flotation or special outfitting teaches respect for the river very quickly. Minor mishaps were common: almost every trip included a pinning or two and everybody got to lend a hand. Concepts like leadership and skills like rope handling were thus part of everyone's introduction to the sport, and when something serious happened everyone was fully trained and ready to respond. Contrast this with the experience of today's paddlers: thanks to their superior equipment and training, they may become involved with Class IV to V whitewater without ever having seen a serious pinning. Problems in such difficult rivers are often severe, but if paddlers have only a limited background to draw on their response may not be equal to the challenge . . . unless the group has made an effort to acquire the needed skills elsewhere and in advance.

While the paddling public has become less involved with rescues, professional outfitters have refined their safety-and-rescue skills to a high level. Their intensive training, combined with the sheer number of inexperienced people they encounter, makes them the people to consult on the subject. The authors of this book, Les Bechdel and Slim Ray, have been leaders in this area for some time. Bechdel, a former national kayak champion and now vice president of the Nantahala Outdoor Center, won the Red Cross Certificate of Merit for his actions in one of the rescues described in this book's early chapters. Slim Ray, a talented author and photographer also associated with the Nantahala Outdoor Center, blends Bechdel's hard-earned wisdom with his own experience to produce a superior text. As a result, this book will be of great value to the advanced paddler concerned with developing rescue skills.

But remember: reading this book is not enough! If you wait until trouble strikes before you practice, as we did in the old days, you may find yourself with more than you can handle. Take the time to develop these skills, either through informal training sessions or through clinics offered by established canoe clubs and professional outfitters. It will increase your appreciation of the river and give you the confidence you need to tackle difficult rapids. And the added margin of safety will make your sport more enjoyable.

PROLOGUE: TWO RESCUES

A Drowning on the Chattooga
by Slim Ray

It was a trip like any other. After a high-water year the Chattooga had finally settled down to a medium level (about 1.8 feet), and we had started to relax a little. There were six rafts and a safety kayak on the trip. Rick Bernard, an expert paddler, had traded a raft-guide slot on another trip to be safety boater. It was a fine November day and we did not expect trouble. There were no problems moving through the Five Falls area and when we reached Jawbone, the next-to-last rapid of the series, we waited for Rick to run through first. He decided to catch the eddy above Decapitation Rock, a huge undercut boulder about halfway through the rapid. This was common practice for safety boats and Rick had done it before. The danger of the move is that this steep, small eddy flows out directly into the undercut end of the rock.

Rick caught the other eddies in the rapid nicely, but hit the eddy above "Decap" low and began to slip back. We had seen this too; the boater must flip here deliberately to avoid hitting the rock with his head. As Rick slid under the rock we watched the downstream side to see him roll. Nothing happened.

The next thing we saw was his hand reaching up from beneath the rock to try to find a hold. An alert guide threw a rope, but Rick was unable to hang on to it. After a few more seconds the hand disappeared. Shortly after that his lifejacket washed out.

Rick's boat was pinned about a foot from the stern and folded underneath the rock. It was made mostly of nylon and other synthetic fibers that would not tear (a breakaway cockpit would have been of no use in this situation anyway), and the problem was compounded by the lack of footbraces. Since the boat was now facing downstream, the force of the current was pushing Rick against the front of the cockpit rim.

There were people on that side of the river, including one of our rafts, but communication was difficult because of the noise of the water. It took several long minutes for them to understand

3

exactly what the situation was. Even then, the rescue attempt was hampered by inadequate equipment. Rick's boat was barely visible from the top of the rock, and the whole force of the current slammed into the upstream side of the rock. We had ropes, but nothing else, not even a carabiner, to hook onto the boat. The boat's grab loops were out of reach, and although I was able to reach and release his sprayskirt in the hope that the boat would fill up and wash off, I couldn't get a rope around the boat. We tried every method we could think of for over an hour, as hope for Rick's life faded.

Dave Perrin, the trip leader, had to think about the rest of the people on the trip. It was getting late and we were overdue. The guides ran the rafts through Jawbone and met the customers below. Dave appointed a substitute leader and the trip set off for the take-out. At the same time he sent a runner out to notify our company of the situation. Since we expected help, Dave and I stayed there.

Just before dark Payson Kennedy (director of the Nantahala Outdoor Center), Andrew Stultz, Bruce Hare, and a member of the local search-and-rescue outfit showed up. With fresh people and ideas we renewed our efforts. We didn't want to leave our friend under that rock! Finally, Andrew formed a coil of rope and dropped it in above the rock right where Rick's boat had gone in. The water sucked the rope in just as it had the boat, and the rope looped around the folded part of the boat. In the gathering darkness we secured the rope to the boat and passed it over to the other shore. Pulling directly against the current, it was all six of us could do to pull it out. It had taken us five hours to find the right combination for the recovery.

Rescue on the Bio-Bio

by Les Bechdel

In January 1981 three commercial rafting outfitters were making their way through the infamous Nirreco Canyon of the Bio-Bio River in Chile. Our group was scouting a quarter-mile-long Class V rapid called Lava South and had pretty much concluded it was too big for our paddle rafts. After constant rain the river was running over 20,000 muddy cubic feet per second and things looked ominous.

With another outfitter, who was camped below the rapids, we watched the third outfitter's rafts crash their way through Lava South. The size of the rapid made control marginal, and their rafts were barely making it into the eddy above the next rapid, Cyclops (another Class V).

As we watched I saw an emergency signal from one of our guides at the top of the rapid. A raft was coming through upside down and one of the passengers was backstroking weakly for the left shore. We lost sight of him for what seemed an eternity in the big waves. When he flushed out in the tail waves he was floating face down.

Three of us — all employees of the Nantahala Outdoor Center: Dick Eustis, Drew Hammond, and I — were on a 15-foot cliff overlooking the fast current that led into Cyclops Rapid. We all saw what had happened and each of us reacted differently: Dick dived into the water immediately, I followed, and Drew, who was further downstream, began getting a throw rope ready. A boatman from the other company was already on his way to his oar rig to back up the rescue. Each of us acted on instinct, based on his position relative to the victim. There was no time for discussion, only action.

Dick swam to the victim and tried to give mouth-to-mouth resuscitation while they were both still in the water, but he wasn't successful in the big waves. I reached the victim and we started swimming him in, using a sidestroke with each of us holding a lapel of his lifejacket to keep him on his back.

Drew made a remarkable rope throw, but it was a long one and only Dick was able to grab the end. The force of the current on the three of us was too strong, so I let go and swam for it. Drew did a dynamic pendulum belay, running downstream as Dick and the victim swung across the eddy line.

John, a boatman from the other company, was there in his oar rig and pulled the victim out of the water onto a large dry box. A quick check confirmed that he was not breathing. Dick started mouth-to-mouth resuscitation and Drew began chest compressions. In their haste they were out of sync, but John started counting and got things together for effective resuscitation.

His friends across the river couldn't help us but shouted that his name was Billy. Within a few minutes Billy started to vomit and we had to roll him on his side. I checked his pulse at the carotid artery to measure the effectiveness of the compressions and arranged for a backup CPR team.

After a few minutes of this Billy blinked his eyes and moved. We stopped CPR, he moaned, and by God he was breathing on his own! He started shivering violently but was still unconscious. We cautioned each other that we might have to restart CPR. We moved him from the raft to a tent with a caterpillar pass and began warming him under sleeping bags with our bodies. He regained consciousness that night and was able to walk the next morning.

In retrospect, it is fortunate that we were in the right place at the right time. But without preparation, skill, and organization, that luck wouldn't have mattered.

Two stories. A drowning in one and a successful rescue in another. How and why did they happen? These narratives give a first-hand account of what happens in a whitewater emergency. You may not understand now exactly what went on or why, but the purpose of the rest of the book is to try to answer those questions.

Whitewater boating is a challenging sport and the rewards are great, but the inherent risk of injury or death must be recognized. Too much that has been written about whitewater either ignores this danger entirely or overdramatizes it to impress the reader. We have tried not to do either. The safety of any adventure sport is directly related to the knowledge of and preparation for its hazards. You will better enjoy your sport if you are prepared to deal with its dangers.

Introduction

Compared to other adventure sports, like skiing and mountaineering, whitewater river sports are relatively new. Their history in the United States really begins just after World War II, when a small group of entrepreneurs began taking people down the Colorado River in rafts made from war-surplus bridge pontoons. Today's boom began in the 1960s, when new designs, technology, and materials made possible the boats and inflatable rafts we know now and opened up new rivers for exploration and recreation.

As with any new adventure sport, there were accidents and fatalities. River runners in general, and kayakers in particular, soon earned a reputation as daredevils, a reputation many of them cheerfully accepted. And there certainly was a great deal of danger on those early trips. Some of the danger was due to the primitive equipment of the time, but more of it was due to ignorance of the dangers of the river and of the potential of rescue techniques.

As the sport's popularity increased in the 1970s many articles and books were written about it, but these concerned themselves mainly with paddling technique and trip reports. In spite of a growing number of accidents, the literature of safety and rescue remained sparse. It is to fill this need that *River Rescue* was written.

Some of the techniques mentioned in this book are simply the application of common sense; others have been learned through experience, both useful and bitter; and some are adaptations of proven mountaineering techniques. We want to emphasize two things: first, that these techniques are still evolving and will continue to change and develop; second, that it is important to share experiences and techniques with others. There is no denying a certain Southeastern bias in this book, since it is based on our own experiences, but we encourage input from all areas of the country. A problem that arises in discussing whitewater rescue today is the lack of a central clearinghouse for statistics, accounts of drownings and near misses, and new safety and rescue techniques. All readers are encouraged to send accident reports, photos, or clippings of rescue- and safety-related material to the River Safety Task Force of the American Canoe Association, Box 248, Lorton, Virginia 22079. This information is compiled annually and made available to paddlers nationwide.

Realizing that rescue systems and concepts of safety differ from one area of the country to another, we have tried to incorporate only systems and techniques we have tried out and found to be effective. This doesn't mean that if a technique isn't in here it doesn't work or shouldn't be used; just that we've used the ones we've described and know that they work.

This is not a "how-to-paddle" or first aid book. We assume that the reader has a basic knowledge of whitewater paddling and knows the rudiments of first aid. We strongly urge all paddlers to take a first aid course; it is essential for anyone who works or plays on a river.

We should mention that while both of us as river professionals have a strong and continuing interest in river rescue and safety, it was mainly the death of our friend and fellow guide, Rick Bernard, on the Chattooga River in 1979 that spurred us to write this book.

Finally, we want to thank our employer, the Nantahala Outdoor Center, for its generous help in preparing this book and for its work in advancing the cause of river safety.

L.B., S.R.
Wesser, North Carolina
Fall 1984

The face of the river, in time, became a wonderful book . . . which told its mind to me without reserve, delivering its most cherished secrets as clearly as if it uttered them with a voice. And it was not a book to be read once and thrown aside, for it had a new story to tell every day.

MARK TWAIN, *LIFE ON THE MISSISSIPPI*

· 1 ·
River Sense

When something goes wrong on the river there seem to be two types of people: those who have foreseen the trouble and are already taking corrective steps — and the rest of us, who are standing there with jaws agape trying to figure out what's going on. The first type of person always seems to be just where he's needed at the critical time. Such people have "river sense."

People are not born with river sense; it's something they develop over time. It involves perfecting skills, understanding equipment, and appreciating the forces and hazards of whitewater. It involves an ability to evaluate other people's paddling skill, a sense of group dynamics, and effective communication with other people. It involves simply being alert at all times on the river.

River sense means accident prevention. Most river accidents and drownings are the result of a combination of poor planning, improper equipment, and plain ignorance. In the spring of 1984, for example, five rafters drowned after attempting to run a low-head dam on the Potomac River. Evidently they had no idea of the power of the deadly hydraulic at the base of the dam. Accidents of this nature can be prevented, and they are doubly tragic because they lead to unnecessary regulation by well-meaning public officials who often don't understand that there are better, less restrictive ways to save lives.

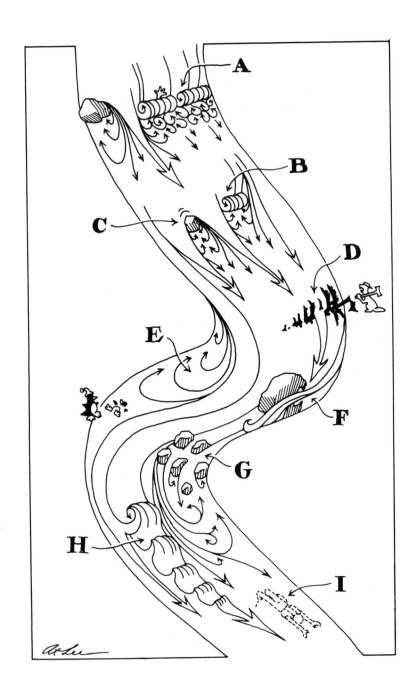

Characteristics of Whitewater Rivers

There are four factors to be considered when assessing the difficulty of a whitewater river: elevation loss, volume of flow, geomorphic make-up of the river bed, and hazards. Other factors, like weather and accessibility, must also be considered but do not strictly speaking have to do with the difficulty of the river.

Elevation loss in the United States is normally expressed in vertical feet per mile, usually as an average figure over the section normally run: the net vertical drop between two points one mile apart. Averages can be deceptive, however, since some rivers concentrate their elevation loss in one steep section, thus hiding a difficult section with easy water before and after. For example, Section IV of the Chattooga has an average drop of about 50 feet per mile, but in the quarter-mile of the Five Falls it averages well over 200 feet per mile. Rivers like this are called pool-and-drop rivers. Other rivers have a steadier elevation loss. These rivers can be more dangerous, since there may be no calm pools in which to recover after a difficult section. Furthermore, there may be the risk of a long swim in cold water if things go wrong. Continuous Class III water may thus be more dangerous than pool-and-drop Class IV.

Volume of flow must be considered alongside elevation loss when determining the difficulty of a river. Flow is measured in cubic feet per second (cfs): the volume of water that passes a given point on the riverbank in one second. Elevation loss and volume of flow are often inversely proportional: a big river like the Colorado will carry over 20,000 cfs (it ran over 100,000 cfs in the spring of 1983) through the Grand Canyon, yet the average drop is just over 8 feet per mile. Some of the scarier creek runs in the East drop almost 300 feet per mile but typically carry a scant 200 to 300 cfs. In general, the higher the numbers, the hairier the paddling. Right now the outer limits of navigability are represented by rivers that combine big continuous water with lots of elevation loss.

The river bed itself partly determines the difficulty of the run.

Fig. 1.1. River features and hazards: (A) Ledge hydraulic, (B) hole, (C) eddy with upstream pillow, (D) strainer, (E) bank or shore eddy, (F) undercut rock, (G) boulder sieve, (H) standing waves, (I) submerged undercut rocks.

Fig. 1.2. A rescue at Bull Sluice. The rope thrower has thrown half the rope to the rafters and the other to a swimmer further downstream.

Narrow river beds may have tight turns and constrictions and are more likely to be blocked by fallen trees, rock slides, or an intrusion of boulders from a side canyon. The geomorphic make-up of the river bed will largely determine the presence or absence of undercut rocks, boulder sieves, potholes, and ledges. All create unseen hazards. Two examples of rivers with extreme geomorphic hazards are the Chattooga in South Carolina/Georgia and the Gauley in West Virginia.

The nature of the gorge through which a river flows also dictates the commitment a paddler must make. Paddling a river in a steep-walled canyon is a much more serious undertaking than trying one with a road alongside.

Hazards

A whitewater hazard is any obstacle or condition that is capable of harming a boater. It may be a hard object like an undercut rock

Fig. 1.3. A geomorphic hazard— the Chattooga's infamous Bull Sluice pothole at very low water. Hazards like this are difficult to see at higher water. A paddler drowned here in 1982.

or a fluid one like a hydraulic. Most whitewater paddling can be looked at as the avoidance of hazards. To avoid them you should understand what they are and how they work.

Holes The novice is often surprised to learn that not all surface water in a river flows downstream. Powerful upstream currents and waves in the form of eddies, hydraulics, and "holes" can be serious hazards under certain conditions.

An eddy is an upstream current that forms behind a surface object in the river or behind a riverbank. As water flows around the object it piles up on the upstream side and then flows inward behind the object, creating a reverse current. The line between the upstream and downstream currents is the eddy line. Larger and faster flows produce a marked difference in height between the upstream and downstream currents. Eddy lines in large, swift rivers can themselves be a hazard as bad as any you may find on the river: they may be several feet wide and studded with cross-currents and whirlpools.

Water also forms a reverse current when it flows over a sub-merged object such as a ledge or boulder, creating one of the most enjoyable but dangerous features of the river, the hole. There are many words for it (pourover, hydraulic, vertical eddy, stopper, reversal, sousehole), but in this book we'll use "hole" to mean the general phenomenon of a reverse current that tends to trap and hold a buoyant object.

Small holes are great fun to play in with a decked boat. Boaters love to see who can go into the biggest one, stay in the longest, and

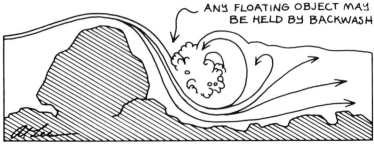

ANY FLOATING OBJECT MAY BE HELD BY BACKWASH

Fig. 1.4. A typical hole caused by water flowing over a rock.

Fig. 1.5. *Some holes are great fun to play in, like this one on Section III of the Chattooga . . .*

Fig. 1.6. *. . . and some aren't, like this one at Sock-em-Dog in the Chattooga's Five Falls. Note the helmet of the boater swimming out just downstream of the boat.*

do the most tricks. But there are a few things to remember: don't stay in until you're exhausted, because getting out is harder than getting in. And look downstream before you go in: what's down there if you have to come out of your boat?

Large holes can be deadly and are capable of holding boats and boaters for extended periods of time. Smooth ledges with no breaks in them and low-head dams form the worst holes. This type of hole is often called a hydraulic. A hydraulic is frequently

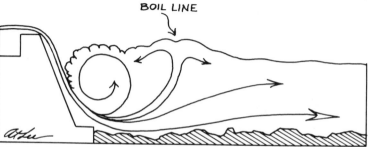

Fig. 1.7. *A hydraulic caused by a low-head dam.*

hard to see from upstream, and the regular nature of the back-wash makes it nearly impossible to get out without help.

On bigger rivers, breaking waves can also form holes as the tops of the waves fall back upstream. Large ones are quite capable of flipping rafts and giving boats a thrashing, but they are generally less dangerous than hydraulics, since an upside-down raft, a swamped boat, or a person will usually flush through. However, mishaps like this often set the stage for worse things on a large, cold, continuous river.

Undercuts and Potholes   It's sometimes sobering to see a river at low water and realize what you have been paddling over when the water was high. The geological reasons for undercuts (large rocks that are narrower at the bottom than the top) and potholes (smooth, eroded depressions in rocks; sometimes the rock will be worn right through to form a tunnel) are not important to the paddler, but their existence and location are. At higher water undercuts can often be recognized by the lack of an upstream pillow of water, and frequently they will have water boiling up behind them. The danger of an undercut is that a boat or a person can be pushed under it and trapped by the force of the water, pinning of the boat, or entrapment of an extremity. Undercuts also collect logs and other river debris, which form strainers and increase the risk of entrapment. The most dangerous undercuts are the ones on or near your line through a rapid, the ones the current pushes you directly into.

Entrapment   One of the biggest dangers of an unplanned swim in whitewater is entrapment, a general term for getting any body extremity, usually a foot or a leg, caught against the river bottom by the force of the current. Often a person unfamiliar with whitewater will attempt to stand on the river bottom and walk to shore. This is an invitation to get a foot caught in a tapered crack between two rocks or in an undercut ledge. Once the limb is caught, it is held there by the force of the current. Escape is difficult and it is likely that the person will drown. If the river is fairly deep (more than about 4 feet), the chance of foot entrap-ment is smaller, unless the drops are very steep — in vertical or

Fig. 1.8. Another common hazard on some rivers is the undercut. An undercut is a likely pinning spot and should be avoided. (Photo by Robert Harrison/Whetstone Photography)

Fig. 1.9. Entrapment.

Fig. 1.10. Left Crack of Crack-in-the-Rock Rapid is just wide enough to wedge a paddler's body in and hold it with the force of the water. The danger level is somewhat higher than the level in this photo.

near vertical drops a swimmer is forced from a horizontal position to a more or less vertical one, which increases the risk of entrapment in a boulder sieve or rock crevice at the bottom of the drop.

Left Crack of Crack-in-the-Rock Rapid on the Chattooga is a good example of a place where entrapment is a danger. The opening is very narrow and tapers down to less than the width of a person's body at the bottom. At higher water the crack fills in and the water will carry a swimmer over the deadly tapered section; at very low levels there is not enough water to carry a swimmer into it; at medium levels it is a killer: the water carries the victim directly into the taper and wedges him there with the full force of the current.

Strainers   Fallen trees and accumulations of debris can form strainers: water flows through but a solid object like a boat or person won't. A strainer can completely block a narrow river. Strainers are particularly dangerous because they look so innocent. Frequently they are also hard to see, especially when a fallen tree trunk is stripped of branches and partly submerged. Not only trees are dangerous, though: smooth, water-worn boulders pushed down from a side creek can form boulder sieves at low water and trap boats and people or, at high water, form a series of holes.

Debris   In addition to all the hazards nature has put in our way, there are man-made ones to contend with. Especially in the East, many rivers have all sorts of debris left over from the days of mills, dams, and logging that can pin or spear a boater. Bridge pilings are particularly likely to pin rafts or boats, since they have little if any upstream water cushion.

Low-Head Dams   Low-head dams and weirs deserve special mention, since they have caused a number of drownings. They are common in the East and usually appear as a uniform feature all the way across the river. They form the perfect hydraulic: regular, difficult to see from upstream, and almost impossible to get out of without help.

Fig. 1.11. A strainer out in the main current on the Nantahala. Unlike fixed hazards such as low-head dams and undercuts, strainers can be a surprise, since they can be created or move overnight.

Fig. 1.12. A low-head dam.

An important feature by which to recognize a hydraulic at the base of a low-head dam is the boil line. This marks the boundary between the water flowing back upstream and the downstream flow. The water in the backwash is white, frothy, and aerated, while the water flowing downstream is darker and smooth. At the line where the two meet, the water appears to be boiling up. *Anything that gets further upstream than the boil line will be pulled into the hydraulic by the backwash.* Rescuers must keep this in mind. Several firemen and search-and-rescue personnel who were not aware of these hazards have drowned while attempting to rescue people from hydraulics below low-head dams. One woman rescued from the backwash of a low-head dam commented that she had been through much worse-*looking* rapids. With good reason low-head dams are called "drowning machines."

This all points to the need for good educational preparation. Know what you're looking for and looking at.

Fig. 1.13. Another potentially lethal hazard is the low-head dam, which can form a regular hydraulic across the entire river. Each year there are several deaths in these "drowning machines."

Big Water Even without the dangers of hydraulics and holes, big and continuous whitewater is a hazard for a person out of a boat or raft. Breathing, no matter how good your lifejacket, is difficult, and so is any kind of rescue. Swimming in this kind of water can lead to "flush drowning."

Cold Water Long, cold swims quickly bring on hypothermia, a lowering of the body's core temperature that robs the swimmer of his strength, judgment, and, if prolonged, his life (see Appendix I).

Rating the Rapids

Interpreting the classifications of river rapids causes a lot of confusion. The American Whitewater Affiliation (AWA) International Scale of River Difficulty grades rapids in Classes I to VI, from "easy" to "extreme risk of life." The "Grand Canyon" system, or "Deseret scale," sometimes used in the West, rates rapids on a scale from 1 to 10. The problem with any of these scales, though, is that they attempt to define a fluid phenomenon with an arbitrary number. One person's Class III is another's Class IV. Pad-

Fig. 1.14. What class rapid is it? Present classification systems are vague and should be applied with caution.

Fig. 1.15. Education is an important part of paddling. Here students at one of the Nantahala Outdoor Center's rescue clinics practice a single-boat Telfer lower.

dlers are getting better, and by today's standards older guidebooks often overrate rapids. Nantahala Falls, for example, which used to be rated Class V at high water, is now considered by many to be an easy Class III. Some Western boaters have taken to downgrading rapids as a matter of course. A Class VI which has been successfully run becomes a Class V and, after a few more runs, a Class IV.

The result of all this is a system in chaos, which in some ways is worse than no system at all. Several changes, such as an open-ended scale similar to that used by rock climbers or a decimal subgrouping within each category, have been proposed but not adopted. Right now the only way to be sure how difficult a river is is to compare notes with other paddlers. Compare the new river to one you've already paddled: maybe it's harder than Noname Creek but easier than the Bottomdrop River. Get specifics: don't settle for vague descriptions like "It's only Class IV" or "It's easy." And no scale should be a substitute for your own judgment.

Preparation

Adequate, functioning equipment is critical for safety. Equipment preparation includes jobs like patching boats and repairing lifejackets and helmets, tedious work best done in the off season rather than the night before the first spring trip. It also means having the right equipment and having it *with* you.

Some paddlers go to great lengths and cost to prepare their equipment but do very little to prepare themselves either physically or mentally. Knowing your own physical abilities and limitations is important. Equally important is knowing that they change with the seasons. Boating is a seasonal sport, and after a winter's layover few of us are at the same peak paddling ability as we were last fall. It's better to warm up on a few easy rivers than to begin pushing the limit the first day out. Your chances of injury (not to mention muscle soreness) are much less if you've maintained a regular fitness routine over the winter.

Time spent off the river can be used to expand your abilities and horizons — the middle of a rapid is not the best place to practice your strokes. If your roll is weak, get in a pool or on a lake and sharpen it up. Many clubs sponsor weekly pool rolling sessions during the winter months. Education in general is an important part of paddling. Practice and expand your paddling skills in the controlled environment of a clinic. Take a class in rescue techniques, first aid, or cardiopulmonary resuscitation (CPR) so you don't have to guess what to do in an emergency. Learn from books, magazine articles, club slide shows, outfitters' presentations, and paddling guides.

Pre-Trip Considerations Before you get on the water gather information about where you're going. A thorough knowledge of the river is essential, and a remote creek requires more research than a popular river run.

Consider the river. Where will you put in and take out? Are there any alternatives? How long is the run, and can it be completed in the time available? Add some extra time for lost shuttles, playing, and emergencies, and remember the days are shorter in the spring and fall. What is the level of difficulty of the river? How much will it change if the water goes up? Where are the river gauges (and what are the safe levels?), the major rapids and hazards, and emergency-evacuation access points and trails? What about the location and phone numbers of the nearest rescue squads and hospitals? Are there any landowner problems? Save yourself the pleasure of looking up the wrong end of an irate farmer's shotgun and find out before you leave. Are guidebooks or USGS topographical maps available for the river and the surrounding area?

Screen participants for paddling ability. It is much easier for a trip organizer to say no to an unqualified friend over the phone than at the water's edge, but judging paddlers you have never met or seen paddle is difficult. Ask some discreet questions about rivers they have paddled or the names of others they have paddled with to get an idea.

Arrange for shuttles. There is no better way to get any trip off to a bad start than by a poorly organized shuttle. A botched shuttle can compromise the safety of the trip if it puts you on the

Fig. 1.16. Check before you leave. Outfitters' shops can often provide information about the rivers you want to paddle, as well as be a source of last-minute gear and provisions.

river much later than expected and forces the group to bend safety rules to hurry. Some forethought about meeting locations, finding drivers, avoiding trespass on private property, and learning about road conditions will contribute to a timely start. A good map is a big help, and the mechanical condition of the shuttle vehicles must be considered. Almost everyone has a story about the shuttle that got lost, got stuck, or broke down.

Have a contingency plan. What happens if the water is too high or too low? or if the river is dam-controlled and there is no release? Can other sections of the river be run? Are there other rivers in the area of the same level of difficulty? Finding out river levels and local weather conditions is an art. Driving all night to get to a river only to find it in flood invites rash acts.

On the Water   Once at the river, get together and discuss the trip so that everyone knows what the plan is. Talk about the length of the trip and the nature of the river and compare notes about the weather and water temperature to help determine what to wear.

Choose a leader. Paddlers tend to be independent and informal and sometimes resist this, but in the organizing stage of a trip and especially in an emergency there are good reasons for having a leader. A leaderless group may be disorganized in an emergency, and that is the very time instant action and coordination are needed. Small paddling groups (three to five people) of nearly equal ability seldom formally choose a trip leader, but in that case each individual must accept responsibility for his own actions and be prepared to assist other group members in any rescue function. In informal groups like this, individual initiative must substitute for leadership.

Designate a lead and sweep boat, so that experienced paddlers will be first and last on the river. In larger groups (ten or more boats), two or three boats can stay together as "buddy boats" the whole day. This spreads the impact of a large group over more of the river. Major rapids and a description of the take-out should be noted before starting. Times should be agreed on, so as to keep the group moving during the day and avoid getting caught out after dark.

Someone should be assigned to carry a first aid kit and a repair

Fig. 1.17. Organization on the river is important. Paddlers should stay within sight of each other, and everyone should keep an eye on the boats ahead of and behind them.

kit, and everyone should know which boats these are in. In general, the first aid kit should be carried by the most medically competent person in the group and should stay near the sweep boat. Groups of more than ten boaters should consider having more than one first aid kit.

Fig. 1.18. Sometimes horizon lines aren't easy to see from a boat. This one conceals a 12-foot drop on the Nantahala. Scout what you can't see.

Fig. 1.19. Scout as a group. If you're not confident of your ability, portaging is an honorable option.

Scouting Scouting is looking before you paddle. It is the last step in preparation before actually going into a rapid. This is normally done in unfamiliar or difficult rapids, but it's always a good idea, even if you've run the rapid before. If the water has been high, for example, there might be a strainer down across the river.

When do you scout? A ledge or vertical drop will appear first as a "horizon line" across the river. The noise of the water may be louder and mist and spray apparent. Difficult rapids often appear in areas of geologic change, which you can spot by watching the shoreline. Do the contours of the land drop suddenly? Is there a staircase appearance to the boulders on shore or in the treetops? Is the river narrowing down and the banks rising? All these are invitations to scout.

Some people are able to scout almost everything in their boat, eddy by eddy. Most can't, and the best way is to walk. If possible, scout from both shores. You'll be amazed how the perspective changes. If possible, scout from river level: if you look down on the river from above, things flatten out and don't look as big as they really are. Obvious routes seen from on high have a way of getting lost in a confusion of waves and holes once you are on the water. Pick markers to designate your line: a particular boulder, a wave, or a landmark on shore will help define your route.

An experienced paddler will carefully study the rapids and plan the run almost stroke by stroke. However, an all-too-common attitude among decked-boat paddlers is "If I miss my line and flip, I'll just roll up at the bottom." Such nonchalance has got some people into trouble. Consider the consequences if things go wrong and have a contingency plan ready. Where are the eddies? If you end up swimming, which way will you go? How close is the next rapid?

Group scouting is usually best, because it allows for an exchange of opinions and gives the trip leader a chance to suggest tactfully that some paddlers should walk. Peer-group pressure can be a source of trouble. There is nothing wrong with running something a little over your head, as long as adequate safety precautions are taken, but there are those who will encourage you to run something just to see you get hammered. If you don't feel up to it, portaging is an honorable option.

Leadership

Paddling is an individualist's sport, and often leadership and teamwork are purposely avoided. Yet organization, whether for a trip or for a rescue, is vital and requires leadership. This doesn't mean giving orders, except in an emergency, but designating a member of the group to give directions may make the difference between a well-organized trip or rescue and a disaster. A good trip leader is a person with experience and river sense. He should have good judgment and good enough interpersonal skills to get along with the rest of the group, and he should be skilled on the river. "Getting along" might mean the ability to organize things without appearing to give orders, or to politely say no to an unqualified paddler. Good leadership may mean walking around a rapid you'd like to run yourself in order to encourage a weaker paddler not to run it, or it may mean keeping your eye on someone who has taken repeated swims and might be becoming hypothermic. It means being a clearinghouse of information for the other trip members, and it means always checking *everything*. The trip leader is not always the strongest paddler in the group, but he should be someone with a cool head and the ability to make decisions.

Saying No

We have already mentioned that you need to measure your own ability against the river. There will also be times that you'll have to measure the abilities of others, even strangers, and say yes or no to them. This requires a good eye for paddling ability and a little diplomacy. At one extreme are those who invite novices along just to see them get munched; at the other are those who refuse to let anyone but the best come because they do not want to take time out from boating for teaching or assistance.

Experts sometimes seem either to lose perspective about the difficulty of paddling a particular rapid or deliberately play it down in order to make themselves look better. This makes it hard for others to gauge the difficulty of rapids objectively. Add to this the vagueness of the present rapids-classification system and you have a situation in which it's easy to get in over your head. If

Fig. 1.20. Leadership on the river can be critical during a rescue. Even small groups should consider electing a leader.

measuring your own skills against an uncertain standard seems hard, it is even harder to measure someone else's. Ultimately it is the individual paddler who determines his qualifications, but the trip leader and the other paddlers should express candid opinions on the matter. Rather than asking if a prospective group member can paddle Class IV water, the leader should ask what rivers he has run and at what water levels. Once at the put-in, some practice rolls or braces will tell the good leader more about the person's ability than any amount of verbal description.

The "What If?" Factor

Good river sense demands a special type of awareness. As you paddle you are tuned in, often on a subconscious level, to your own performance and that of other members of the group. You are aware of the hazards of a particular rapid and are always ready to begin a rescue if needed. Thinking like this works well when scouting rapids. As you scout, pick out the hazards and potential trouble spots. Are there ways of avoiding or minimizing the hazards? If there is a flip or a swim, what will happen? What rescues could be used? In the back of your mind you are always asking yourself "What if?"

At first you will have to think consciously about the "what if?" factor, but after some practice it will become automatic. Your mind will be continually assessing potential problems, discarding some alternatives and working out the details of others. Some may criticize this as pessimistic paddling, but by avoiding accidents and being ready for trouble we are able to be more positive about paddling in general. The test comes when something goes wrong. Time for rescue is measured in minutes and sometimes seconds. The "what if?" factor may be the difference between success and failure.

Conclusion

We close this section with an account by Charlie Walbridge of a drowning on the South Fork of the Clearwater River in Idaho (full details may be found in the *Best of the River Safety Task Force Newsletter, 1976-1982*, published by the American Canoeing Association in 1983).

The kayaker in question was clearly in over his head, having lost a boat on another, easier river just two days before. He flipped, swam, and drowned in the first rapid of the run. Walbridge notes that the victim "had a tendency to 'follow' people, feeling that, in or out of his boat, he'd probably make it."

Trusting in probability is a poor substitute for good judgment and a realistic assessment of one's own abilities in any case, but what about the other group members? "Many in the group felt," says Walbridge, "that someone should have said something to Chuck." There does not seem to have been a designated trip leader, but on the other hand "no one ... expected him to attempt the river that day; he jumped in the water at the last minute." Chuck's miscalculations were his own, but other people in the group were aware of them and did not make their reservations known.

"What's that metal thing on your lifejacket for?"
"I don't know, but it must be important because
everyone else has one."

CONVERSATION OVERHEARD ABOUT A
CARABINER, OCOEE RIVER, JUNE 1981

· 2 ·
Equipment

On the river, proper equipment is as important as good judgment. We have seen too many trips ruined by improper or missing equipment, and shoddy equipment has been linked to a number of accidents. Unfortunately, equipment problems can affect everyone on a trip, not just the individual paddler.

The general heading "equipment" includes personal safety gear the paddler wears; safety equipment like ropes and throw bags, which are usually carried separately; and the watercraft he paddles, whether a raft, a canoe, or a kayak. The choice of equipment is determined by the difficulty of the water, the time of year, and the location of the river. Obviously a first descent in the Andes in spring and an afternoon float on the Buffalo River will require different equipment. But the minimum, *always*, is a lifejacket for everyone and a helmet for those in decked boats.

For some people, the preparation is almost as much fun as the paddling. We see them outfitting boats and patching gear on long winter evenings. Some are equipment freaks, with dozens of gadgets they'll never use; there are others whose idea of getting ready for the season is to buy a new roll of duct tape (these are usually the ones who have forgotten something important, need a shuttle, and just remembered they cracked their paddle last fall). On the river you should be able to concentrate on paddling and not have to worry about equipment failure. You must choose your

Fig. 2.1. The well-equipped kayak. This one has a tow system installed behind the cockpit, a butane lighter and small emergency kit in a film canister in the substantial rear wall, and a throw bag clipped into the seat. A prusik loop run through the seat completes the outfitting.

own equipment, but there are some specific criteria everyone should consider.

Personal Safety Equipment

Lifejackets The U.S. Coast Guard calls these personal flotation devices (PFDs) and with typical organizational mania has classified them into Types I through V. We only need to consider Types III and V.

The primary purpose of the lifejacket is quite simple: to help keep your head above water so you can breathe. (We say "help" because in some water no lifejacket will keep your head above water all the time.) Some compromises must always be made between flotation, comfort, and mobility. This partly explains the difference between Type III and Type V jackets.

Type IIIs are cut for comfort and mobility and have the minimum amount of flotation needed for safety. The best ones are cut in vest fashion, with a zipper in front, and have sticks of expanded

Fig. 2.2. A National Park Service ranger models an outfitted lifejacket during a rescue course. Note the whistle on the zipper, knife on the waistband, carabiner on the shoulder, and another carabiner hooking the prusik loops together at the waist.

PVC sewn in. The amount of flotation varies with the manufacturer, and some ("shorties") are cut very short to provide improved mobility for kayakers with a minimum of flotation. In general, the bigger the water, the more flotation is required of the lifejacket.

Type Vs were designed with commercial use (i.e., whitewater rafting) in mind. They have more flotation and a collar to help turn the unconscious wearer face up (fine for a lake but of little use in whitewater). Most Type Vs are uncomfortable for paddlers to wear.

Lifejackets have benefits besides flotation, too. The foam will protect you in a fall while scouting or cushion your back if you're swept into a rock either in or out of your boat. A snug, well-fitting lifejacket is also warm: it provides over an inch of insulation for the critical trunk area of the body. "Shorty" lifejackets don't do well in either of these areas.

None of this is going to matter if you don't have your lifejacket on, whether because you didn't wear it or because it came off when you most needed it. A tattered, faded lifejacket with shrunken foam is considered by some to be the emblem of the true river veteran. Forget that image and retire "ol' faithful" for safety's sake. A good test to see if the time has come is to jump in the water, holding onto your lifejacket so that it stays in place. If you end up floating with your nose underwater, it's time. Then try jumping in without holding onto it. If the lifejacket comes up over your head, it's time for some work on the tie straps. A snug, tight fit with a sturdy waist tie is essential.

A river-guide friend of ours is an expert paddler with many first descents (including Niagara Gorge) to his credit. He used to be a real cheapskate when it came to personal gear. His lifejacket had long ago lost its waist tie and his helmet was carved out to fit his huge head. While guiding on the Ocoee one day he was flipped backwards out of his raft and hit his head. "I was helpless," he said later. "I was a competitive swimmer in college but there I was, floating face down, unable to take a stroke." Fortunately, another raft was able to get to him before things got worse, and the first thing he did after getting out of the hospital was buy a new lifejacket.

Fig. 2.3. Some of the many commercial lifejackets available. These are all Type IIIs. From left to right: the "shorty," a larger model with extra flotation, and two "normal-length" ones. (Photo by Ciro Pena/Nantahala Outdoor Center)

There are some ways you might want to modify your lifejacket. One is to add crotch straps to keep the lifejacket from coming off over your head, but this can be a bother if you have to take your lifejacket off and put it on a lot. Another is to substitute a better waist tie for the standard nylon tape supplied with most lifejackets. We suggest one-inch tubular nylon webbing or a pre-tied prusik loop secured with a carabiner.

Helmets These are a must for those using decked boats. We recommend them on difficult rivers for rafters and canoeists, too. If you use a helmet, get one worth having. Some sold today give little more than a false sense of security. Ideally, a helmet should protect not only the top of your head but also your temples and the back of your head. Some paddlers even go so far as to use helmets with football-style faceguards.

Fig. 2.4. Two types of helmet. The one on the left is fiberglass and gives considerably more protection than the plastic one on the right. (Photo by Ciro Pena/Nantahala Outdoor Center)

Fig. 2.5. Three paddling essentials: the carabiner (locking on the left, non-locking on the right), whistle, and knife (folding and fixed). All these are usually carried on the lifejacket for quick access. (Photo by Ciro Pena/Nantahala Outdoor Center)

Carabiners   These are usually associated with rock climbers and mountaineers, but paddlers find them useful for many purposes. The carabiner is truly a multi-purpose river tool: it can secure gear in boats, substitute for a pulley in many situations, attach haul lines to pinned boats, and do a multitude of other things. It is a good idea for everyone to carry at least one and preferably two: complicated rescues sometimes need a lot of carabiners.

There are many different kinds of carabiners. For river use, get an aluminum-alloy one: it won't rust. Don't buy one with a locking gate, because the gate can get clogged with sand. If you need extra security, you can use two normal carabiners clipped in so that they open in opposite directions instead of a locking carabiner. D-shaped carabiners are usually stronger than oval ones.

Most paddlers like to wear their carabiners on their lifejackets, attached at the shoulder or waist. However, you can injure your shoulder or collarbone if you flip over in a decked boat and the carabiner gets between you and a rock. Probably the neatest solution is to have the carabiner clip together two prusik loops and act as a waistband.

The Prusik   A mountaineering knot, the prusik was designed to cinch on a haul line when under tension. When the tension is relaxed, however, the knot can be adjusted. It is made from a loop of small-diameter line tied together with a double fisherman's knot (see Appendix D). The best kind of line for this is a soft 5- to 7-mm kernmantle-lay nylon rope. Outfitters that cater to rock climbers stock this kind of line.

Boaters often call the loops as well as the knot a prusik. We wear two short loops girth-hitched around the waist and clipped together by a carabiner (Fig. 2.6). It's an invaluable piece of equipment for quick tie-offs, anchor ties, hitching systems, and Z–drag rescues.

Knives   These are recommended for recreational paddlers and essential for the river professional. A knife can cut through the floor of a broached raft to free someone. It can cut rope. Some of the haul-line systems we will describe in later chapters can develop tremendous loads when pulling on a pinned boat. If some-

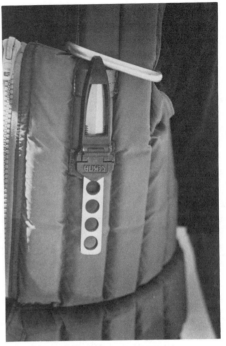

Fig. 2.6. This paddler wears two prusik loops girth-hitched around his waist and joined with a carabiner.

Fig. 2.7. A knife with a plastic locking sheath and a carabiner at the ready. The knife is carried upside down for a quick draw with either hand.

one's arm or leg gets snarled in the line a sharp knife will get him free. One day on the Nantahala Les came upon a raft towing a boy on an inner tube. The raft hit a rock and the boy was dumped into the water. When he came up, his head was caught in a noose of parachute cord. A submerged tree limb speared the inner tube and pulled the cord tight around his neck. "I leaped out of my canoe," Les remembers, "brandishing my knife like Errol Flynn. I cut the cord and swam the kid to shore. The rafters never noticed he was missing. I decided right there that my silly-looking knife was worth every penny."

Get a sturdy knife with a stainless blade and a positive locking sheath that can be operated with one hand. Skin-diving knives work well but tend to be rather bulky. Many paddlers use folding knives, but they are hard to open with one hand and are therefore not such a good choice. A serrated edge is a good feature: you might have to saw through a boat. Keep your knife handy: strap or sew it on the chest, shoulder, or waist of your lifejacket. Some people feel knives look too aggressive and wear them inside their lifejackets, but if you have a harness or rope around you it can make the knife hard to get at.

Whistles   Communication can sometimes be a problem on the river because of the roar of the water. A good whistle can help. For the same reason, you should also know the AWA River Signals (see Appendix B). A whistle is also handy if someone gets lost walking out.

First Aid Kit   The contents of your first aid kit will vary greatly according to the location, season, duration, and anticipated difficulty of any trip. For popular rivers near well-traveled roads you need no more than a simple "ouch pouch." Extended day trips in remote locations with a group require a more comprehensive kit, one able to deal with serious lacerations, broken limbs, illnesses, and insect-sting reactions (anaphylactic shock). Wilderness expeditions require things like suture kits and tooth-fracture treatments, as well as an array of prescription drugs available only through a doctor. (The contents of first aid kits are discussed in more detail in Appendix F.)

Emergency Kit   Often carried with the first aid kit, this is intended to cope with emergencies other than physical injuries. A butane lighter or a fire-starting kit is a good idea if there is a chance you might be caught out overnight or need to warm a hypothermia victim. You might carry an extra set of contact lenses if they are essential, or a small space blanket. Some paddlers keep items like this in a waterproof container sewn into their lifejacket.

Clothing

Proper clothing is as important as any other piece of equipment. In the past few years new synthetics such as pile and polypropylene have made substantial inroads into the wool and neoprene that have dominated the boating scene since the 1950s. Whatever you wear, the object is the same: to maintain the body's core temperature. In order to do this, certain critical areas, the trunk and head especially, must be protected against heat loss.

On the river the problem is that the paddler is constantly wet and may at any time be immersed in very cold water. The water acts in two ways: the body is wet by waves or spray, the water evaporates in the wind, and the body is rapidly chilled; or, when the body is immersed in cold water, there is rapid and severe heat loss by convection, which leads to immersion hypothermia in a very short time.

Outer Layer The first defense is an outer layer that will protect the body from the chilling effect of the wind. Usually this consists of a paddling jacket and pants made of some kind of coated nylon fabric to keep spray and waves off the bare skin. Both jacket and pants should be cut loose enough to fit over the insulating layers underneath.

Insulating Layer Wool, nylon, or polypropylene pile or knits all make excellent insulators under a paddling jacket. In milder conditions they can be worn alone. The same layering principle applies here as in mountaineering: two or three thin layers are better than one thick one. Avoid cotton clothing.

The new synthetics are so good that they have led some paddlers to advocate eliminating wetsuits and drysuits altogether. This can be a serious mistake, particularly for novices, because neither synthetic materials nor wool will protect you in the water. Water will flow right through any fabric and directly over the skin, causing an abrupt and massive heat loss to the body. This may be an acceptable risk for experts, but if you're still learning and expect to swim a lot, keep reading to the wetsuit and drysuit sections.

Fig. 2.8. To stay warm you need at least two layers — an outer layer to cut the spray and an inner layer of pile, wool, or neoprene. (Photo by Ciro Pena/Nantahala Outdoor Center)

Fig. 2.9. Neoprene is still the best material for extreme conditions. Wetsuits come in all shapes, sizes, and thicknesses. (Photo by Ciro Pena/Nantahala Outdoor Center)

Wetsuits   Capable of replacing both the insulating layer and the outer layer, the wetsuit was developed for skin divers. Made of neoprene and coming in a variety of thicknesses and coverages, it works by letting a small amount of water in between the skin and the neoprene. The body warms this thin layer of water, which then acts just like any other insulating layer. If the wetsuit fits properly (i.e., tightly), this layer of warm water will be held against the skin even when the wearer is in the water. The neoprene will also, of course, act as a shield against waves and spray.

Against these advantages, however, it must be said that wetsuits are constricting to the active paddler and somewhat uncomfortable. "Plumbing" arrangements are inconvenient for men and next to impossible for women. Wetsuits also do not lend themselves well to layering, though the new generation of thinner wetsuits is much better on this count. The fact is, though, that wetsuits are still the best protection in extreme conditions.

When deciding whether or not to wear a wetsuit, use the "hundred-degree rule": if the combined temperature of the air and water is less than 100°F, wear a wetsuit.

Drysuits A relatively recent development in whitewater clothing is the sealed, watertight suit, or "drysuit." The openings at the wrist, ankle, and neck are watertight and the suit is worn over normal insulating clothing. With water sealed out and the body kept dry, heat loss is minimal; in fact, heat build-up is sometimes a problem. Drysuits will undoubtedly become more popular.

Throw Ropes and Throw Bags

The safety rope is the primary tool of rescue. Used by itself as a throw rope or with carabiners and pulleys to retrieve pinned boats and rafts, it forms the basis of many rescues. Every paddler should carry one.

Ropes are found on the river in two main forms: the throw bag and the coiled polypropylene rope. A throw bag has the rope stuffed rather than coiled inside a nylon bag, along with a chunk of flotation foam. The advantage of a throw rope is that only the required length need be thrown — someone in a hole or headed for an undercut doesn't need 65 feet of rope unspooling in there with him, which is what can happen if you use a throw bag. If the rope is divided into two coils, each coil can be thrown to a different swimmer.

What features should you look for in a rescue rope? Ropes for whitewater use should float and be visible on the surface of the water. The ropes we use are made of yellow polypropylene, which meets both these requirements. Polypropylene rope is superior to natural-fiber ropes in that it does not absorb water and to nylon ropes in that it does not sink. The rope should be large enough to be comfortable to work with and strong enough for mechani-

Fig. 2.10. The "standard" river-rescue rope is a 60- to 70-foot length of ½-inch braided yellow polypropylene.

Fig. 2.11. Les models a new development in throw-bag technology — the baby bag. Carried at the paddler's waist for short throws, it's especially handy for rafters in big water.

Fig. 2.12. Throw bags are more easily carried and deployed than ropes but have certain limitations. Most bags use the 3/8-inch polypropylene rope shown here. (Photo by Ciro Pena/Nantahala Outdoor Center)

cally assisted rescues. For throw ropes, we normally use a 60- to 70-foot length of rather stiff braided or twisted ½-inch polypropylene, with a breaking strength of about 4200 lbs. Throw bags, on the other hand, normally use a similar length of a softer lay of ⅜-inch polypropylene with a tensile strength of 1750 to 2200 lbs. The rope should be long enough to reach a swimmer but not so long that it can't be thrown effectively.

Safety ropes must be stored in such a way as to be easily accessible and readily deployed. Because of their size and stiffness, standard ropes are difficult to store and slow to deploy. For this reason most paddlers use throw bags, but while a throw bag is more easily stowed and deployed, it is difficult to restuff or coil quickly after the initial throw. Also, because of the "bucketing" effect of the bag in the water, which makes it difficult to retrieve the rope for a second throw, a standard rope is much quicker for repeated throws. The bag is also more easily snagged while being retrieved.

There have been some new ideas in throw bags. One is a quick-release chest harness with a small throw bag attached to the back, which you wear over your lifejacket. If you are entrapped, you can throw the bag to rescuers and be pulled to safety. (The quick release is in case something goes wrong.) Another idea is the "throw sock." This uses a bag of twice the normal length with a 100-foot line in it. The extra length of the bag gives some added leverage to throw with. Still another idea is to use open-weave nylon (fishnet) fabric for the bag to reduce the "bucketing" effect of the bag in the water.

Boats and Rafts

Whitewater river craft can be divided into three broad categories: rafts, open canoes, and decked boats. All have benefited over the years from improved designs and technology, and these improvements now allow us to extend the limits of the sport to ever higher levels. In extreme whitewater, however, the consequences of mistakes or equipment failure are often severe. To reduce the risk of accidents, it is important to maximize the safety features of each type of craft.

Rafts When selecting a raft, look for multiple air chambers (four or more), quality materials such as neoprene and hypalon, and heavy-duty construction. There should be an adequate number of D-rings attached to the sides of the raft for your purposes, and they should be large and securely attached. Hand lines along the sides should fit snugly to prevent accidental entanglement of the arms or legs.

Two of the most important factors in determining what class of water the raft can handle are the size of the raft and the load it is carrying. Size means not only the length and width of the raft, but also the diameter of the tubes and upturn of the ends, all of which affect the amount of water taken on. Bigger is not always better: maneuverability must be taken into account on many rivers, and although a larger raft may be slower to fill up than a smaller one, it may be so heavy when full as to be unmanageable. Some rafts with larger-diameter tubes have self-bailing floors. The maximum safe load also varies acording to the class of water you expect. Do not base your calculations on the manufacturer's figures from the raft's data plate. These are for calm water *only*.

Rafts must always be prepared for a flip. "Flip lines," pioneered by Western rafters, are pre-rigged lines attached at one end to D-rings on the side of the raft or to a rowing frame. With the raft upside down, a boatman can grab the line, stand on the opposite tube, and flip the raft over in mid-current.

All too often, gear is left strewn about the raft floor. This invites losing it in a mishap, so everything should be tied or clipped securely into the raft before you start, and it should be kept that way. In an oar rig, beware of shoehorning yourself into the rowing compartment, because you will need space if the downstream oar gets caught. In one incident, an oar was driven completely through a boatman's thigh because of a crowded rowing compartment.

On big rivers, consider outfitting rafts for double or triple rigging. When two or three rafts are tied together, the combined mass is less likely to be flipped or stopped by a hole. To prepare for this, install additional D-rings low on the tubes so that the rafts can be tied together top and bottom, which will prevent them from flipping over onto each other. The rig is controlled by sweep oars in the bow and stern.

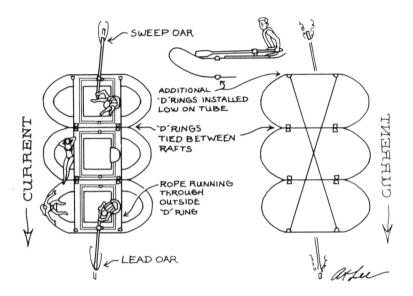

Fig. 2.13. A triple-rigged raft.

Decked Boats   Proper outfitting of a decked boat is important for safe paddling. "Outfitting" means finishing a boat and fitting it to yourself, after you buy it but before you paddle it. For a kit boat this may include seaming, installing foam walls and grab loops, and a multitude of other things; other boats may come almost ready to paddle. However, all boats need a certain amount of modification so that the paddler can fit snugly in the boat. The key here is to be tight enough to "wear" the boat yet not be wedged in so tightly that a quick exit is difficult.

Foam walls of some sort are an important safety item for any decked boat. Mounted vertically under the bow and stern, parallel to the length of the boat, they add deck-to-hull strength: without them the deck is likely to collapse in heavy water or in a pin. Walls are usually made from ethafoam, though styrofoam and minicell are also sometimes used. However, the ideal length and thickness of walls is a controversial topic. The harder the water and the more playing you do, the bigger and stronger the

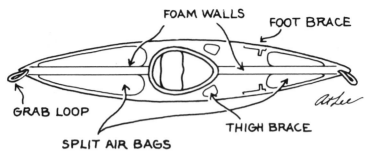

FOAM WALLS   FOOT BRACE

GRAB LOOP

SPLIT AIR BAGS

THIGH BRACE

Fig. 2.14. A safely outfitted boat should have sturdy, braced walls, grab loops, airbags, foot braces, and thigh braces.

walls must be. Airbags may be adequate for Class I and II rivers, but strong walls are definitely needed for harder water. Two-inch ethafoam walls, even with airbags on either side, can fall over sideways or bend under pressure, even if they are braced from the side. Three- or 4-inch walls are much more stable. You should secure all walls, particularly those in plastic boats, by gluing side blocks to the deck and hull to brace the walls at the top and bottom.

Making the walls longer and thicker makes it harder to get out of the boat, particularly for long-legged paddlers, but proponents of these longer walls point out that they keep the paddler from being jammed up to his armpits in a vertical pin. On the whole, the balance of opinion seems to favor longer and thicker walls.

The construction of the boat itself can have important safety implications. Fiberglass boats should have a "breakaway zone" of fiberglass without synthetic fibers around the cockpit area. The fiberglass, which is much more brittle than the tough synthetics of the hull, will fracture if the boat wraps, giving the paddler a chance to escape. This type of construction has saved several lives. Plastic kayaks are not at present made with breakaway cockpits, which has led many to regard them as more dangerous than fiberglass boats. One paddler we know has fashioned a "roll cage" of tubular aluminum for the cockpit area of his plastic boat. It seems to work, but it adds weight to an already heavy boat.

Another important safety item is the grab loops. These should be securely attached at bow and stern and made of at least 8-mm rope, because if you are vertically pinned one of them is going to be the most convenient handle. Some boaters attach a third grab-loop, or "Blackadar handle" (named after the well-known American paddler, Walt Blackadar, who invented it), just behind the cockpit to give a swimmer something to hang on to in big water.

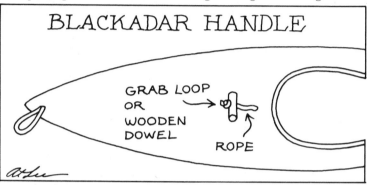

Fig. 2.15. A "Blackadar handle."

A useful feature to add is a towing system for getting swamped boats back to the shore. Most paddlers rely on the "bulldozer" method: putting their bow in the cockpit of the swamped boat and pushing it to shore. A more efficient way is to use a towing system that puts you in a normal ferry position upstream of the swamped craft. Any tow system should have a foolproof release system, unless you like the idea of having a 1000-lb sea anchor permanently attached to your boat. The jam cleat, a standard sailing item available through most marine hardware stores, works very well. Be sure to test the release on any tow system before you use it.

Any of the several tow systems shown here (Fig. 2.18) will work well. Most are inexpensive and easy to install. The center-mounted system (aft of the cockpit) allows the tow boat to control the ferry angle precisely. Systems that tow the boat near the grab loop make ferrying more awkward but are easier to attach initially.

Open Canoes   The new generation of ABS designs has allowed open canoes to run water once reserved for decked boats. Even the Colorado River in the Grand Canyon has been run in an open canoe.

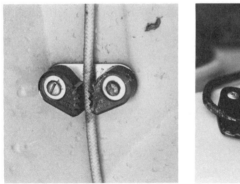

Figs. 2.16A and B. The heart of the tow system is the jam cleat. Two types are shown here. Either may be set for towing and then quickly released.

Fig. 2.17. The tow system installed on a Dancer. It can be used for towing boats, carrying retrieved paddles, and ferrying lines.

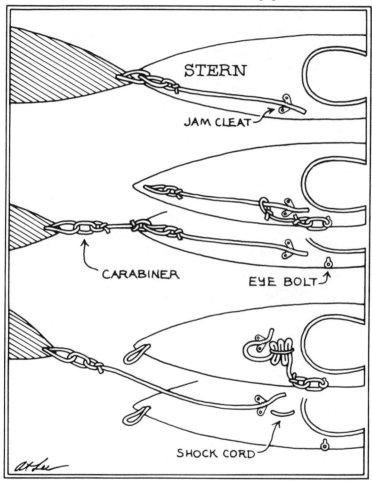

STERN

JAM CLEAT

CARABINER

EYE BOLT

SHOCK CORD

Fig. 2.18. Three types of tow system.

Some people choose open canoes because of the reduced risk of entrapment. Walls are not necessary, as they are in decked boats, but a fitted styrofoam block placed under the thwarts (Fig. 2.19) will still go far to eliminate the danger of a canoe pin or paddler entrapment. Blocks are bulky, which leads some people to prefer airbags, but while airbags will displace water and make the boat

Fig. 2.19. This canoe has a fitted styrofoam block in the center. It takes up room but makes the canoe less vulnerable to pinning.

easier to swim to shore they do not add any structural strength. Another system of flotation, especially popular in New England, involves attaching a 4-inch-thick layer of ethafoam to both sides of the hull of the canoe, which gives you more room to carry duffel.

Canoeists often install thighstraps or braces to improve their control over the boat. Straps should be made of a wide, stiff material that won't snag the paddler in a wet exit. Position them so that they don't come up around your groin but fit lower down, toward your knees. Velcro is the best strap fastener, since unlike a buckle it will release under strong pressure.

Unless you line your canoe a lot, there is little reason to use painters over 8 feet long. A loose painter becomes a hazard in an upset and may snarl the paddler. Store it under a shock cord when not in use. If you have aluminum gunnels, go over them periodically with a file and remove any sharp edges. A spare paddle is a good idea, but make sure the hold-down system works: you should be able to get the paddle easily but have it secure if the boat overturns.

Conclusion

Many safety decisions are best left to the individual. Others, such as lifejackets and helmets in decked boats, should be universally recognized as necessary. Decisions such as whether to wear a knife or install a towing system should be determined by the difficulty of the water and your personal preferences. There is a definite attraction to paddling with a minimum amount of gear, and each of us must balance the rewards with the consequences.

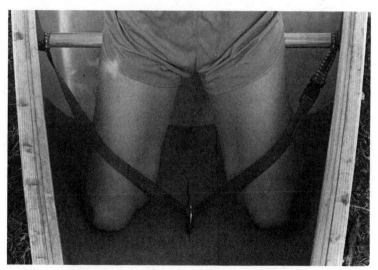

Fig. 2.20. Thighstraps are an integral part of the outfitting of most whitewater canoes. They should allow the paddler to exit without entanglement. This set has a quick release on the paddler's left. (Photo by Ciro Pena/Nantahala Outdoor Center)

Fig. 3.1. For those in decked boats the quickest means of self-rescue is the Eskimo roll, but it must be practiced to be effective.

God helps them that help themselves.
BENJAMIN FRANKLIN, *POOR RICHARD'S ALMANAC*

· 3 ·
Self-Rescue

Rivers keep flowing, regardless of our personal misfortunes. An important lesson for all paddlers to learn from this is that you have to take care of yourself on the river. Charlie Walbridge, of the River Safety Task Force, says that most of the recreational paddlers who drowned in American rivers in 1983 were not entrapped and were wearing lifejackets. They perished floating through whitewater, unable to get to shore.

In Chapter 1 we saw the importance of preparation, both mental and physical, and good judgment. We also discussed some of the leader's responsibilities for the safety of the group and the individual. In this chapter we'll concentrate on the individual paddler, and on how he can deal on his own behalf with some of the hazards we've defined.

We recommend that you don't paddle alone, but even in a group the reponsibility for your own safety rests in large part with you. You are the one who wanted to paddle, and you are the one who must ultimately take the blame for your miscalculations. It follows that you should be able to rescue yourself.

For those in decked boats, the quickest and most obvious means of self-rescue is the Eskimo roll. It is the best insurance policy you can invest in, and like any other aspect of the sport it requires practice. A good roll is a real boost to your self-confidence, which in turn reduces the likelihood of more serious mistakes. Nevertheless, there is also a time to swim. Overconfidence and misplaced pride ("I never come out of my boat") are as bad as too little confidence. In open water it is all right to keep trying to roll, and if you scouted thoroughly you should know when you turn over if there is a hazard coming up. Often it's a matter of experience: novices tend to bail out too soon and experts to stay in too long. Experience will give you almost a sixth sense for where you are in a rapid when you find yourself upside down.

Paddlers in open canoes, most of whom do not have the luxury of a roll, must rely on the swimming technique of self-rescue. When that last low brace fails and you capsize, get to the upstream side of the boat when you come up. If things look bad, abandon the boat and swim for shore. If it seems safe to do so, work your way to the upstream end of the boat, grab the painter (but don't wrap it around your arm), and try to swim the boat over to shore. Sometimes it is possible to right a swamped boat and paddle it to shore.

Rafters can also use self-rescue after a flip. If the raft is equipped with flip lines, as described in Chapter 2, it can often be righted while in the water. If the raft can't be righted, the rafters should attempt to climb on top of the capsized raft and paddle it to shore, after first checking to see what is downstream. Another method is to have a line or throw bag rigged so that a crew member can grab it, swim to shore, and pendulum the raft in. If this is not possible the crew should try to swim the raft to shore by holding on to the D-rings or painters.

Swimming with a boat or raft can be more dangerous than swimming without one, especially on rocky rivers, where there is a chance of getting squashed between the boat and a rock. In big water the extra buoyancy of your boat is an asset, but stay on the upstream side. Wide, cold rivers, such as the ones in Canada and Alaska, present another problem. Although the whitewater may not be difficult, a swimmer may die of hypothermia before reaching shore. With a swamped decked boat a paddler can sometimes reduce his exposure by crawling on top of the boat and paddling it like a surfboard. Pairs of open-canoe paddlers in this situation can try a flatwater technique: the Capistrano flip (Fig. 3.8). This is done by getting under an upside-down canoe, then doing a scis-

Fig. 3.2. The pre-packaged flip line is a new development in rafting self-rescue. It stows neatly out of the way, but after a flip . . .

Fig. 3.3. . . . it can be pulled out and used to right the raft as shown.

Fig. 3.4. Another method of self-rescue for the rafter is to get up on the upside-down raft and paddle it to shore. Obviously this only works in fairly calm water.

Fig. 3.5. Yet another method of rafting self-rescue. A throw bag is tied and clipped into the bow D-ring of the raft . . .

Fig. 3.6. . . . and the rafter unclips the bag after the flip, swims to an eddy, and pendulums the raft in.

sors kick and flipping the boat at the same time. Both canoeists then grab the gunnels, and while one holds on the other slides into the canoe with a scissors kick. This cannot usually be done if there is duffel in the canoe.

It's always a good idea, if you can, to hang on to your equipment if you have to swim. However, bear in mind that equipment is cheap and lives are not. This may sound obvious, but at least one paddler drowned on the Chattooga because she chose to hang on to her boat rather than a throw rope and was entrapped in the infamous Left Crack.

We recommend that every paddler swim a "safe" rapid at least once a season to renew his respect for the force of the water and to keep in practice if his roll fails. Novices and experts alike should do this. Charlie Walbridge notes in the *Best of the River Safety Task Force Newsletter* that "many experts make lousy victims." Often the transition from hero paddler to helpless victim is a hard one.

The standard defensive position for swimming in a rapid is lying on your back with your feet downstream. On shallow, rocky rivers try to maintain a horizontal position as near the surface as possible, always looking downstream and preparing to fend yourself off rocks with your feet. Don't float passively from one rapid to the next: kick with your legs and backstroke with your arms. Swim aggressively and use boating techniques like ferries to let eddies help you move toward the shore. In deep water with big waves, a crawl stroke on your front will work better than swimming on your back. Head for shore as fast as possible, especially in cold water. Rescue yourself; *never assume someone else will do it.*

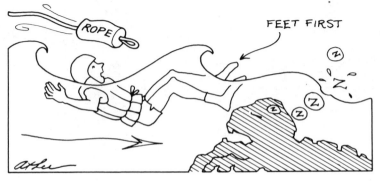

Fig. 3.7. Defensive swimming.

Fig. 3.8. Though not normally considered a method of whitewater self-rescue, the Capistrano flip can be used by canoeists on wide, cold rivers where the whitewater is not too severe.

Fig. 3.9. Two canoeists swimming in whitewater. They are upstream of the boat, facing downstream with their feet up and in front of them.

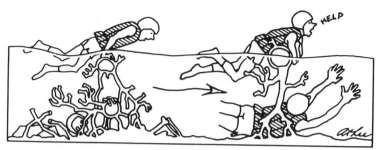

Fig. 3.10. Strainers.

Strainers

If you find yourself floating unavoidably toward a tree-choked strainer, change your position from feet first to head first. A feet-first position will tend to make you wash under the trunk or limbs, which is the last thing you want. Try to swim faster than the current, using a crawl stroke, and look at the tree. Is the trunk slippery? Are there limbs you can grab? You must try to get *up* on the tree and out of the current. As you hit the tree, let the force of the current assist in keeping your momentum up as you crawl onto it. If your boat is floating with you, don't let it block your movements or pin you against the tree.

As a last resort if you can't get up on the tree, whether because the current is too fast or the trunk too large, plan to dive underneath it, head first. Time your breathing and plan on pulling yourself through the branches. Don't let your body get parallel to the main trunk.

Entrapment

From the self-rescue standpoint, the best way to deal with entrapment is to avoid it. This means not trying to walk in water deeper than your knees and swimming with your feet up and in front of you. Once an entrapment occurs, it is difficult, if not impossible, for a victim to escape unaided. It is hard to stand if your foot gets caught, and if your whole body is immersed the force of the water will be much greater than on just the entrapped limb. If you are entrapped, though, you may be able to gain valuable rescue time if you can push yourself off the bottom and dog paddle with your hands. If the water is not too swift, you may be able to get an occasional breath and survive until someone can reach you.

Broaches, pins, and boat entrapments will be covered in more detail in Chapter 6, but here we'll consider some things a paddler can do to help himself in these situations. A boat is broached when it is pushed sideways onto a solid object by the current and held there. The danger of broaching is that the boat may collapse and entrap the paddler.

In a broach, the severity of the situation depends on the position of the boat and the amount of water pushing on it. Only experience can give you a real appreciation of this, but if the water pressure is great enough to blow your sprayskirt off the cockpit rim you are in dire danger of wrapping the boat, and it's time to get out in a hurry. You may have a chance to lean into the rock and push yourself off it with your hands. Sometimes you can push off with your paddle or shift your weight so that the boat moves, but remember that the boat may wrap at any time: you must be prepared to get out as quickly as possible. The more doubtful you feel about the situation, the less time you should spend trying to get the boat off. A quick exit is usually the best choice.

Holes

Paddlers in decked boats often surf holes to practice balancing in the hole and then paddling out. They enjoy doing this, but they also know that sometime in the future they will find themselves in a hole by accident and may need those escape skills. Before surfing an unknown hole, look at it carefully. Is water flowing out of the sides? Is there a weak spot in the backwash to escape through? Consider also what is downstream if you have to swim out. Some paddlers experiment by sending in a friend first ("Sure, it's okay. Surfed it last week. Go ahead while I adjust my sprayskirt"). This method has its adherents, but it doesn't win friends.

In a decked boat the best system of self-rescue in a hole is to paddle out. Try different directions: some sections of the backwash may be weaker than others. Paddle forwards and backwards and try drawing downstream. An expert will stroke on both sides

Fig. 3.11. Paddlers in decked boats often surf holes to practice balancing in the hole and then paddling out.

in a kayak, not just on the downstream bracing blade. Expert canoeists should be able to switch and brace on either side. If paddling fails, turning over in the hole and extending the paddle will sometimes allow the downstream current beneath the backwash to catch your body and the paddle and pull you out. As a last resort, pop the sprayskirt and hope the less-buoyant boat will wash out with you still in it. Keep trying to get out, but don't make the mistake of letting yourself become exhausted. In a violent hole it takes a lot of effort to stay in control; if you can't get out, swim for it early. The best escape is to flip upside down and dive down under the backwash.

An open canoe caught in a hydraulic will almost always swamp quickly, which may cause it to wash out. But be careful: being in a large hydraulic with an open canoe can be dangerous if you and the canoe are trading places.

If caught in a hole, rafters should quickly move to the downstream side of the raft to avoid flipping. Sometimes the raft will fill with water and flush out on its own. Sometimes it can be paddled out or roped out from shore, but a raft full of water is *heavy* and may have a will of its own. You may have to swim out and leave the raft in the hole, especially if there is a risk of hypothermia. Get out by diving out over the backwash of the hole from the downstream tube of the raft.

It is very dangerous, if a raft is caught in a hydraulic, for a crew member to end up underneath it. The swimmer will be recirculated in the usual way but will not be able to breathe if he comes up under the raft. Sometimes swimmers in this situation can be reached by hanging over the downstream tube and feeling under the raft.

Low-head dams form a hazard from which escape by self-rescue is very difficult. A swimmer will pop up at the boil line only to find himself sucked back into the water pouring over the dam. The water will then force him down to the bottom and then up again at the boil line, only to have the whole frightening process start over again. To complicate matters, some dams have exposed reinforcing bars of steel, which can entrap or skewer a swimmer. Debris floating around in the hydraulic (logs, for example) can hit a

Fig. 3.12. Rafters caught in a hole on the Ocoee. If the raft flips or someone falls out there is a danger of a swimmer getting caught underneath the raft.

Fig. 3.13. Self-rescue in a low-head dam.

swimmer, though you can sometimes hang on to a piece of debris for additional flotation. If caught in a low-head dam, try to take guarded breaths and stay relaxed. Save your strength: you can't fight the water. Try to work your way toward the shore while being recirculated. More often than not there will be sheer concrete retaining walls buttressing the sides of the dam, but you may be rescued from the sides if someone can reach you. Sometimes, in smaller hydraulics, it is possible to swim out by diving down and catching the underlying jet of water as you are being recirculated. Otherwise, try to swim downstream only at the area of the boil line, preferably just before you surface. *Do not take off your lifejacket* in an attempt to swim out of the hydraulic: if you fail you will be in much worse shape than before.

Sometimes you have to use your imagination. One fireman got out of a low-head-dam hydraulic by dragging himself along the bottom of the river, clinging to exposed reinforcing bars and rocks.

Conclusion

Not all methods of self-rescue will be found in a book. A friend of ours found himself trapped in his boat at the bottom of a drop on Alabama's Little River. He is a big, strong guy and rather than drown he began tearing the cockpit out of his fiberglass kayak with his bare hands. The boat was a commercial model with a breakaway cockpit, and he was able to break enough of it away to get his legs out and swim free. If it's life or death, almost no method is too extreme.

Fig. 4.1. The basic body belay. This rope thrower is ready to belay when the rope comes taut. He should be wearing a lifejacket.

Never travel far without a rope! And one that is long and strong and light . . . . They may be a help in many needs.

J.R.R. TOLKIEN, *THE LORD OF THE RINGS*

· 4 ·
Rescue by Rope

The most useful tool in any rescue, from the simplest capsize to the most complicated Tyrolean rescue, is the safety rope. In the equipment chapter we compared the throw bag and the throw rope. The choice is up to the user, but every whitewater paddler should carry a rope and know how to use it. In this chapter we will cover the uses and techniques of rope rescue and will use the term "rope" to refer to both the throw bag and the safety rope.

The Throwing Rescue

Most rope rescues are made with a rescuer standing on the shore, throwing a rope to a victim, and hauling him in. It sounds simple, but few people can accurately throw a rope—rope throwing requires practice. The goal of a good throw is pinpoint accuracy at the rope's full extension. The position of the safety rope is the first consideration. A common mistake is to stand directly opposite the most likely point of capsize in an effort to get closer to a potential accident. The best place is usually further downstream, because there needs to be time for the paddler to surface and look at the rope thrower before the rope is thrown. How far downstream you position yourself will depend on such factors as the speed of the current, the nearness of the next rapid, how many roll attempts may be made in a decked boat, and the presence of eddies to swing the swimmer into. Remember: if you are too far upstream and miss, you may have a hard time getting back down along the riverbank. It's often better to be a bit too far downstream.

In long or hazardous sections of whitewater, you may need several rope throwers. If possible, they should be positioned above or opposite hazards like large keeper hydraulics and undercut rocks. In sections of continuous whitewater, finding a pool or eddy for a rescue may be a problem.

The thrower must also consider what to do once the victim has got hold of the rope. The current will usually swing him into the shore like a pendulum. Position yourself so that the swimmer doesn't swing into a greater hazard or get caught in the full current. Look for an eddy to land your catch in. This may require you to move after the throw has been made, so check your route before you throw. Many rope rescues have failed because the rescuer had lead feet.

Before throwing, always try to get the attention of the swimmer. Yelling or whistling will usually get him facing the right way, but it is better to establish eye contact. Once the rope is in the water it is hard for the swimmer to see it. Even though the rope floats, currents will often suck it beneath the surface. Some people, especially those new to the river, will become very disoriented while swimming a rapid and will not respond to you. If you can't communicate before throwing you must try to actually hit the swimmer with the rope (usually on the head) to get his attention.

In the excitement of a rescue the throw is often made too soon —while the swimmer is still upstream of the rope thrower's position. This increases the distance the rope must be thrown and so

Fig. 4.2. Setting rope: Position A gives good visibility and has an eddy behind, but is of little use to a swimmer or an upside-down boat in the fast water of the chute; B is a better place to pick up swimmers coming out of the chute or hydraulic, but the rope thrower must get them in before the strainer downstream; C is set close enough to a hole to rescue a swimmer either in or below it. Both C and D are set to swing a swimmer into a convenient eddy; E is set to catch a swimmer who might go into an undercut; F is too close to the undercut for a rope throw, but this would be an effective rescue position if a paddler were trapped there; At G, would-be rescuers must be ready to throw. This rescuer is too close to the hazard and would not be able to hit a swimmer washing out of the hole or a decked boat after several roll attempts; H is a better position: below the hazard with an eddy nearby.

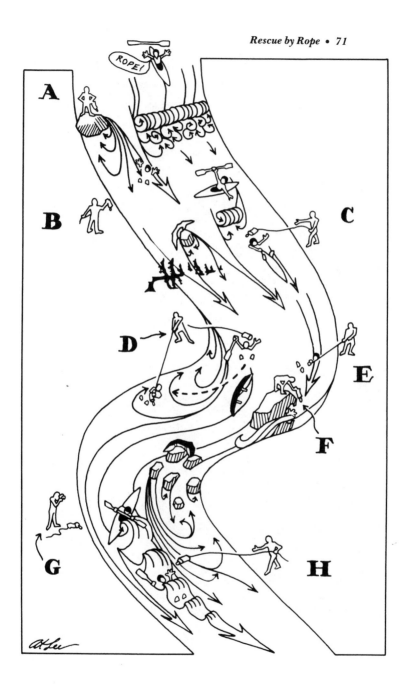

Fig. 4.3. This rope thrower has the rope wrapped around her body in a simple body belay. She is near enough to a tree to be able to wrap the rope around it to increase friction if the body belay isn't enough.

Fig. 4.4. The friction belay uses something like a tree or boulder to increase friction. The rope can still be released if need be.

Fig. 4.5. The dynamic belay. The rope thrower runs down along the bank to decrease the loading shock on the swimmer. The banks must be fairly open to use this technique.

usually decreases the accuracy of the throw. The throw should be timed so that the swimmer is at the minimum possible distance from the thrower.

If you miss, it is better to err to the *upstream* side of the swimmer. The rope will float faster on the surface of the water than a swimmer who is backstroking against the current with his feet pointing downstream, and the rope is likely to drift into the swimmer without having him compromise the defensive swimming position. If the rope is downstream of the swimmer, he will have to swim head first to retrieve it.

Ideally, the rope thrower should stand somewhat above the level of the river. Standing on a boulder or up on the bank gives better visibility and allows more time for the rope to pay out while in the air, but too much height will reduce the working length of the rope and increase the time you must allow for it to reach a swimmer who may be moving downstream very quickly. For most situations, the optimum height is about 4 to 6 feet above the river.

Never tie off the rope when throwing to a swimmer: the rope

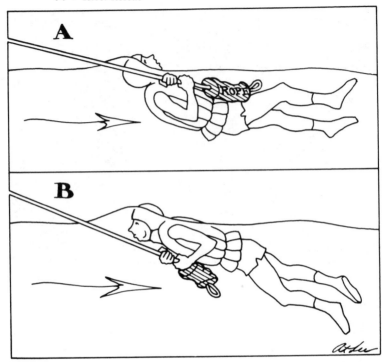

Fig. 4.6. (A) The right way to hold a rope in whitewater. The rope is over the shoulder and held close to the chest, with the swimmer facing up and away from the rope thrower. The swimmer tends to plane to the surface and the water forms an air pocket around his face. (B) The wrong way. The swimmer, lying on his belly and looking at the rope thrower, gets a faceful of water. His body may even tend to dive — the last thing he wants!

may become entangled around a part of his body, even his neck, and you may need to let go of the rope to release the tension. If you think it may be difficult to keep hold of the rope, use the belaying technique instead. Pass the rope around a tree, a boulder, or your own body to increase the friction on it and give it extra holding power (Figs. 4.3–4.4). Belaying is more effective than simply holding the rope in your hands and much safer than tying it off.

Belays may be either dynamic or static. In a static belay, the rescuer throws the rope, assumes the belay position, and "puts the brakes on." The disadvantage of this technique is that when the rope suddenly becomes taut the impact may be too great and the swimmer may lose his grip on the rope. The dynamic belay (Fig. 4.5), which brings the swimmer to a more gradual halt, is often better. The rescuer moves downstream with the swimmer while swinging him into shore. This reduces the loading shock on the swimmer. Another form of dynamic belay is to let the excess rope run through your hands, then gradually tighten your grip on the rope to reduce the braking impact on the swimmer. The rope should be run over your lifejacket, and you must carefully watch the speed of the rope to avoid burns. A long throw will not usually allow enough excess rope for this technique. Dynamic belaying is also a useful technique where there is no nearby eddy and the rescuer must follow the swimmer.

Swimmers should go for the rope aggressively (remember self-rescue? Help yourself!). When you get to it, hold the rope to your chest with your feet downstream. Face upwards and do not look at the rescuer: turning over and facing toward the rope may cause you to dive. Never wrap the rope around any part of your body (in cold water, when your hands may not be functioning well, you may clamp the rope under one armpit and then hold it against your chest). In the proper position the victim will plane on the water, and a breathing pocket will form immediately downstream (Fig. 4.6).

Throwing Techniques

There are three primary methods of throwing a rope or bag: underhand, overhand, and sidearm. Each can be effective, but no matter which style you use it is important to throw with your whole body and not just with your arm. Watch a javelin or discus thrower: they use their whole body, hurling from the legs through the upper torso and following through with the arm. To get the full extension of the rope you must do the same.

Most people stand with their body facing upstream. To make use of your full power potential, however, you should stand sideways, with your throwing arm on the side away from the river. If you are right-handed and are standing on river right (i.e., on the right bank as you face downstream), this means your body will

Fig. 4.7. The correct throwing position for a right-handed thrower — facing slightly downstream, looking over the shoulder.

Fig. 4.8. Dennis Kerrigan shows one way to hold your end of the rope while throwing. Always be sure you can release the rope quickly.

Fig. 4.9. The coiled rope is split into two coils and thrown as shown.

be facing downstream and you will be looking upstream over your left shoulder.

Before throwing, look for obstacles: slalom-gate wires, bystanders, or tree branches. If you are using a throw bag, loosen the cord lock but do not open it to the widest position. Opening it all the way will cause a "bucketing" effect as water flows into the mouth of the bag. When throwing a standard rope, split the coils of the rope between your throwing hand and your holding hand. The loops of rope must be evenly sized and not snarled. If possible, take one or two practice throws to help judge the current and work the kinks out of the rope.

It may seem unnecessary to add that the thrower should hang on to the rope, but we have seen even professionals throw the whole rope away. Be sure you have a good grip on the end before you let fly.

Underhand This is the most popular and perhaps the most natural-feeling throw. The thrower swings the rope back and forth a few times to develop a rhythm and then releases it. The

Fig. 4.10. The same principle is applied to the throw bag, after the initial throw. This saves time because you don't have to stuff the bag if repeated throws are necessary.

release should be at about a 45-degree angle to the surface of the water: a premature release will make the rope hit the water only a few feet in front of the thrower; a late release will make the rope go straight up and then down on the head of the would-be rescuer (much to the amusement of everyone but the swimmer). The underhand is best used for close throws. It can be difficult in brushy or otherwise restrictive areas.

Sidearm   Many people find they can improve their distance with a sidearm throw, but this gain often comes at the expense of accuracy. The sidearm motion, similar to that of a discus thrower, makes it easier to use your whole body. It also requires a fair amount of clear space to throw.

Overhand   The overhand throw is superior to the other techniques in many ways but is seldom used, because it is more difficult to learn. Whereas the other methods need some preparation, the overhand throw can be used on the run with little preparation (a rescuer can run into a throw, like a baseball player). It requires less room to throw and can be used on brushy banks, over the heads of a raft crew, and in waist-deep water. The rope can also be thrown with greater velocity over greater distances. With a coiled rope it is more difficult than the underhand or sidearm throws, but it is ideal with a throw bag.

Fig. 4.11. Position for the underhand throw.

Fig. 4.12. Underhand throw.

Fig. 4.13. Position for the overhand throw.   *Fig. 4.14. Overhand throw.*

Multiple Swimmers

A rescuer is sometimes faced with a situation involving multiple swimmers — when a raft overturns, for example. If the swimmers are bunched close together, throw the rope to the middle of the group. If they are separated, throw to the ones furthest away but still within reach of your rope. With luck, some of the others can grab the rope, too. If you have a throw rope split into two coils as described earlier, you can throw one coil to one nearby swimmer and the other coil to the next. Be sure to hang on to the middle of the rope.

Bear in mind that a single rescuer will have a hard time holding more than two swimmers with a static belay unless he is able to use a fixed object like a tree. In that case, use a dynamic belay, which can handle a surprisingly large load.

Multiple swimmers should all hold on to the rope rather than each other.

Tag-Line Rescues

Unlike throwing rescues, which are intended for moving swimmers, tag-line rescues are used for a victim who is involuntarily fixed in one position — for example, some cases of foot entrapment, a pinned boat, or a person held in a keeper hydraulic.

A tag line is simply a line stretched across a river and brought to the level of a stationary victim. A floating tag line has some sort of flotation device attached to it to keep the rope on the surface of the water and to provide something for the victim to hang on to. A snag tag, on the other hand, is a weighted line used to snag submerged objects.

Tag-line rescues require planning and team effort but have the advantage of being relatively uncomplicated and safe for the rescuers. Untrained spectators can often be used to manage the ropes from shore.

Floating Tag Lines A floating tag line is very useful when a swimmer is caught in a keeper hydraulic like a low-head dam. In a hydraulic like this a person can survive for a surprisingly long time if the water is not too cold and if he does not panic. In many situations like this the victim simply cannot be reached by throw

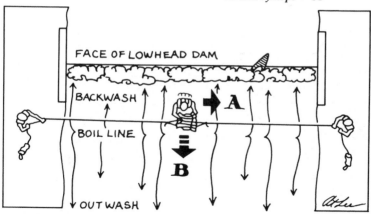

FACE OF LOWHEAD DAM

BACKWASH

BOIL LINE

A

B

OUTWASH

Fig. 4.15. The floating tag line rescue.

rope or by boat. By timing his breathing and staying relaxed (or as relaxed as one can be in such a situation!), the swimmer may gain enough time for a tag-line rescue.

The first step is to get the rope across the river. On a narrow river you can simply throw it across. On wider rivers you may have to ferry the rope across, using whatever craft is available. Before beginning the ferry, tie a buoyant object to the middle of the rope for the victim to grasp. A lifejacket is suitable. Coil the rope neatly on the shore to reduce the chance of entanglement, and position it as high above the water as possible to minimize the drag of the current on the rope (Fig. 4.16). Normally, the rope will be managed by a "rope team" of two or more people.

Try to keep the rope out of the water during the ferry. On a wide river this can't be done, and you should start the ferry far enough upstream to allow for the drag of the current on the rope. The ferrying craft carries only the end of the rope; the paying out process is managed from shore. A kayaker may carry the rope in his teeth or in one hand or tie it to the grab loop or tow system of his kayak. A tandem open canoe can carry a third person in the middle to hold the rope. Rafts are the best for ferries like this, because the rope handler can stand if necessary.

Upon reaching the other side of the river, the ferry team becomes a rope team and, in coordination with the rope team on the other side, moves the buoyant object as required to reach the

Fig. 4.16. The rope ferry. Keep the rope as far out of the water as possible, to minimize drag. This team uses a canoe with a third person to hold the rope high.

Fig. 4.17. A floating tag line. A lifejacket provides the buoyant object for the swimmer to grasp.

victim. The efforts should be coordinated by a rescue leader standing in a position visible to all and directing the action with prearranged hand signals.

Once the swimmer has got hold of the flotation device, he is normally pulled to one side of the hydraulic and rescued. If this isn't possible, both teams can try moving downstream simultaneously, pulling the swimmer with them, to break the grip of the hydraulic.

The tag-line rescue is premised on the victim being conscious and capable of clinging to a floating object. What do you do when the victim is unconscious? The Ohio Department of Natural Resources' Division of Watercraft has developed a flotation ring with unbarbed treble hooks on it to snag an unconscious victim's clothing. The recreational paddler will have to improvise. A method of last resort might be to use a rescuer wearing several lifejackets as the buoyant object. This is a desperate measure, posing great danger to the rescuer, and a rescue team should only consider it if they know the victim is alive and no other method will work.

Snag Tags This variant of the tag line uses a weighted rather than a buoyant line. It is set up the same way as the floating tag line, but its purpose is to reach an object beneath the surface of the water, such as a pinned boat or an entrapped victim. The snag tag does take time to set up, so unless the victim can breathe you should consider it a backup to more quickly arranged methods like the strong-swimmer rescues described later in this chapter.

What should you use for a weighted object? An empty throw bag can be filled with small stones and clipped to the line with a carabiner; a rock with sharp edges (find one in the 25- to 30-lb range) can be tied like a Christmas package (Fig. 4.19); a helmet or ammo box full of rocks will work. Use your imagination.

Controlling the snag tag line takes coordination and a fair amount of physical strength. Each rope team needs a minimum of two people. Say your victim's foot is trapped on the river bed. Position the weighted object approximately 10 feet upstream of the victim, then slacken the rope to allow the object to pass over the victim's head, sink, and land just downstream of him. The force of the current will push the weighted object downstream before it hits bottom. Hitting the victim with it will be the least of your worries.

When the weighted object hits bottom there is a small vibration not unlike a fish nibbling bait. This is when the rope needs to be moved back upstream by the rope teams. We say "moved" because the rope teams do not just pull but actually run quickly upstream.

The idea of this method of rescue is that a conscious victim can grab the submerged rope, but it can be used even with an unconscious entrapped victim. To do this, work the rope up underneath the victim's leg to try and pull his foot free. The rescuers should sink the line on the downstream side of an en-

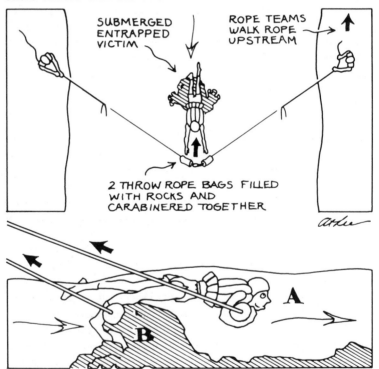

Fig. 4.18. The snag-tag rescue. A conscious victim may be able to grab the weighted object or the rope (A). If the victim is unconscious, the rescuers may be able to work the rope up under his leg to free his foot (B).

Fig. 4.19. Many things can be used for the weight in a snag tag. Here a large rock is tied in like a Christmas package.

Fig. 4.20. The snag tag in action. Here the rescuers are preparing to sink the line. Controlling the line requires a fair amount of strength.

trapped victim or pinned boat and try to "snag" him or it by pulling the line upstream. Rope teams on both sides, directed by the rescue-team leader, should control the line.

Snag tags have also been used to free pinned boats that are totally submerged. The principle is similar to the "rope tricks" discussed in the next section. We will cover pins and extrications in more detail in Chapter 6.

Rope Tricks   How can you *attach* a rope to an inaccessible boat or victim? one which is pinned where there are no convenient eddies or boulders? There are three basic methods: an overhead or vertical rescue, such as a Tyrolean or bridge lower system; a Telfer lower; and "rope tricks." The first two methods, which will be discussed in later chapters, require a fair amount of time, equipment, and knowledge to set up and have definite limitations.

"Rope trick" is a general term that describes methods of using the current to carry a rope around a boat or victim. Once the line is attached, the rescuers can use any shore-based haul system, from a simple direct pull to a Z-drag. Take the example of a kayaker caught in a vertical pin: the boat and the body of the paddler disrupt the current and are frequently out of the water entirely. An appropriate rope trick would be to throw or ferry a rope to the upstream side of the boat and let it float around it. A boat or swimmer could then be used to retrieve the downstream end of the rope and return it to the same shore, thus making a loop of rope around the boat or victim.

In many pins and entrapments the only way to pull a boat or victim out is the way they went in. Similarly, in really inaccessible

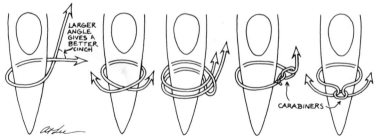

Fig. 4.21. Rope tricks: There are several ways to cinch the rope down on the boat.

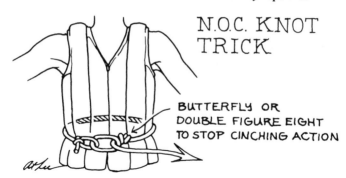

N.O.C. KNOT
TRICK

BUTTERFLY OR
DOUBLE FIGURE EIGHT
TO STOP CINCHING ACTION

Fig. 4.22. The NOC knot trick.

situations the only way to get a rope to them is often to let the same current that put them there carry the rope. (This was the method finally used to get a rope around Rick Bernard's pinned kayak in the incident described in the Prologue, after all other methods had failed.)

Once the loop has been formed, there are several options for cinching down the rope, but the very flexibility of the rope-trick concept prevents our showing every variation. For example, if a boat were vertically pinned really close to shore a paddler with a cowboy background might attempt to throw a lasso over the boat!

If possible, haul on the boat rather than the victim, since even if the line is padded it is likely to injure the latter. If the situation is grave enough, there may be no other choice, though, but to use a cinching rope loop on the victim's trunk area. In such cases, there is a rope trick (the NOC knot trick) which can be used to reduce the likelihood of injury. Tie a knot some 3 to 4 feet from what will be the victim's end of the rope. You can use any type of loop knot— for example, a butterfly or double figure-of-eight knot. Then form a loop, using a carabiner tied to the end of the haul line. As the carabiner slides down the haul line the knot will stop it and create a fixed-diameter loop for the victim. The rescuers should pad the line and attempt to place the loop around the victim's waist. The knot will prevent the line from cinching down and restricting breathing. Injury is still likely using this method, though, and you should not use it unless it is the only way to rescue the victim.

If floating lines can't be used, because the force of the current or the depth of the submerged boat is too great, the rescuers can try a weighted line. You can use an ammo box or throw bag weighted with rocks, just as you did with the snag tag. If the weighted object must slide, clip it to the line with a carabiner.

Strong-Swimmer Rescues

The strong-swimmer rescue is quick to set up and provides some security for the rescuer in the water. By committing himself to the water, however, the rescuer unavoidably increases his chances of injury or drowning. Other, and usually slower, shore- or boat-based rescues should be considered first, if no one's life is in danger. But sometimes there is no alternative: a life-threatening situation, like a paddler entrapped in his boat, demands immediate action, and speed is a major advantage of this method of rescue. An urgent need to rescue the victim may outweigh the potential danger to the rescuer.

For a strong-swimmer rescue you will need a belayer, a rescue swimmer, and a rescue leader, though it can be done with only a belayer and a rescue swimmer. The leader should position himself so that he is visible to both the swimmer and the belayer and can communicate with them using hand signals. The rescue swimmer should have a *loose* loop of rope passed under his armpits. The loop should be loose enough to pass over his head and shoulders if he needs to escape quickly and should be tied with a non-slip knot like a bowline. He should also be equipped with a tightly fitted lifejacket (or even two), a helmet and wetsuit if available, and a knife to cut free of the rope if necessary.

The object of the strong-swimmer rescue is to get the rescue swimmer into the eddy behind the victim. Usually the hazard that caused the accident or the pinned boat itself will create an eddy. In some cases the rescue swimmer can be lowered (i.e., allowed to float with the current directly to the accident site) while in others he will have to swim aggressively to reach it. Once the rescue swimmer has reached the accident scene, he should attempt to

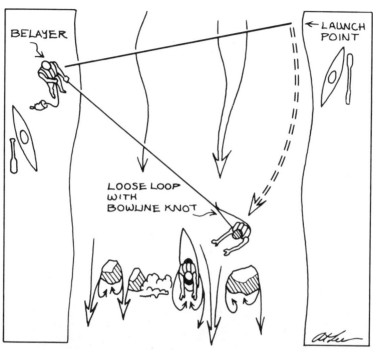

Fig. 4.23. The strong-swimmer rescue: Pendulum approach.

Fig. 4.24. The suggested loop for a strong-swimmer rescue is loose and tied with a bowline with a stopper. The rescuer has plenty of room to slide free of the loop.

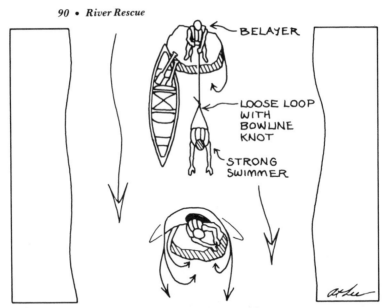

BELAYER

LOOSE LOOP
WITH
BOWLINE
KNOT

STRONG
SWIMMER

Fig. 4.25. The strong-swimmer rescue: Mid-current lower.

Fig. 4.26. The fixed-line rescue. The rescuer is held in the apex of the line by the force of the water but is able to swim free by somersaulting over or under the rope.

free the victim, but his first consideration must be to stabilize the victim's condition and keep matters from worsening. The rescue team could set up a tag line to keep the victim's head above water. They could also set up additional lines (pulled over by the rescue swimmer with his own safety line) to keep the boat from moving. Once the rescuer and the victim are free, they can be swung into shore.

Here are three types of strong-swimmer rescue:

Pendulum Method This rescue is best suited to narrow rivers: the wider the river, the more drag on the belay rope and the more difficult the ferry. The rescue swimmer launches himself from a point upstream of the victim. The belayer, who may be on either side of the river, is usually downstream of the launch point but upstream of the victim. The belayer should be able to see the victim and the rescue swimmer at all times. The pendulum can also be based on mid-river boulders or islands if necessary: protruding boulders between the rescue swimmer and the belayer can cause problems (Fig. 4.23).

Mid-Current Lowers The lowering point is directly upstream of the accident site — on an island, a boulder, a bridge, or a peninsula on a river bend. This method is less risky than the pendulum method.

The Fixed Line The rescue team stretches a fixed line across the river at water level. The rescue swimmer holds the fixed line at his waist and works his way out on the upstream side of it. When the swimmer reaches a point directly upstream of the victim the belayers slacken the line to lower him to the accident side (Fig. 4.27).

As the lowering process continues, the rescue swimmer is held in the apex of the line and his position becomes more and more secure: water tends to cascade around his head, forming an air pocket, and since he is not directly attached to the rope he can escape simply by somersaulting and swimming to safety.

The rope can be ferried across the river by boat or rescue swimmer and should ideally be anchored by belayers on both shores. If there are not enough people the rope can be tied off on

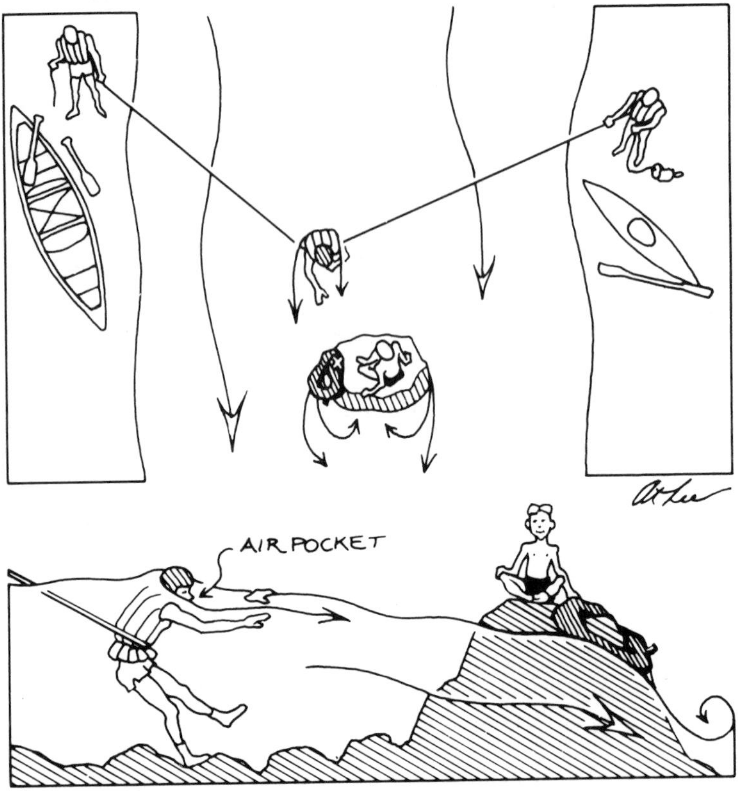

Fig. 4.27. The strong-swimmer rescue: Fixed line.

one side and belayed on the other, but this is more dangerous to the rescue swimmer.

The rescue swimmer can take an auxiliary line out with him if necessary. If he does, someone besides the belayer should hold that line.

Conclusion

No other single rescue tool is as important as the safety rope. It can be used in a wide variety of rescue and evacuation situations. Ropes have their limitations, especially on wider rivers, but all paddlers, guides, and boatmen should make proficiency with a rope an important part of their river skills.

About the only thing I had learned about canoeing was to head into the part of the rapids that seemed to be moving the fastest, where the most white water was.

JAMES DICKEY, *DELIVERANCE*

· 5 ·
Boat-Based Rescue

One of the fastest and most obvious means of rescue is by boat: a waiting boat can be used much more quickly than most shore-based systems. The disadvantage is that it is dangerous to a rescue boater, who must obviously be in the same place as the victim. The harder the water and the less skillful the boater, the lower the chance of success and the greater the chance of ending up with two victims instead of one. A victim clinging to any craft, no matter how cooperative he may be, makes control of the craft more difficult.

Shore- and boat-based rescue techniques are complementary and should be used together. *Where possible,* the primary method should be shore-based, since this is safer for the rescuers, with a boat-based rescue as a backup. Some factors, such as the width of the river, may obviously force a change in the order of preference, but the idea of using different methods at the same time remains valid.

The Eskimo Rescue

Often taught to novices as a preliminary to the Eskimo roll, the Eskimo rescue is one of the first techniques a decked-boat paddler learns, as it provides good practice in staying with the boat. When upside down, instead of attempting to roll, or perhaps after several attempts to roll, the boater leans forward and slaps the sides of the boat to signal that he needs help. A fellow boater

Fig. 5.1. The Eskimo rescue. This is a basic technique of decked-boat rescue.

then moves over and presents the bow of his boat to the sub-merged boater's waiting hands. It is then a simple matter for the upside-down paddler to roll back upright.

The Eskimo rescue has obvious limitations in heavy water. The upside-down boater may float into rocks or into a more danger-ous rapid and draw would-be rescuers in after him. As we said in Chapter 3, coming out of the boat is an honorable alternative. Typically, the Eskimo rescue is used in places like the deep-water tail waves below a rapid or in the eddy just below a hole.

Equipment Retrieval

Retrieving equipment is often a necessary task on the river. In an emergency, though, people must come before equipment, so you should decide in advance who is to go after what. In most cases it is best that shore-based rescuers go after people while boaters go after boats, paddles, and equipment: if the shore-based rescue fails the boaters are still in a position to assist. If numbers permit, some boaters can go after equipment while others back up the primary rescue.

Don't risk people, including yourself, for equipment: if you pick something up, be sure you can unload it quickly if need be. Remember to apply the basic rules of self-preservation: don't run blindly downriver after something without thinking about what lies below.

Paddles Finding lost paddles is easier than carrying them. Paddlers in rafts and open canoes have few problems, but those in decked boats will find extra paddles a real nuisance. They can be laid on the cockpit rim of a decked boat, but this only works in easy water. A C-1 paddler who carries a spare paddle on the back deck can put a retrieved paddle underneath the spare. One very simple method of retrieval is to grab the paddle and throw it toward the shore, repeating the process as many times as necessary. Some paddlers stick the paddle blade down the back of their lifejacket so that the shaft sticks up above their head. We don't recommend this for anything but the easiest water, and then only for skilled paddlers: if the paddler tips over or goes under a low tree-limb the paddle shaft can get caught, which could be very dangerous.

Fig. 5.2. This isn't a recommended way to retrieve paddles. If the paddler tips over he could be in real trouble.

Fig. 5.3. A simple way to recover paddles on narrow rivers is simply to throw them toward the shore.

Fig. 5.4. A more elegant method of recovery is to use the boat's tow system. The blade can be "chicken-winged" under the paddler's arm to hold it, and the paddle can be quickly released in an emergency.

Kayak paddles are even more of a problem, since they're twice as long. One method is to grasp *both* shafts (i.e., yours and the retrieved one) and paddle as if they were one. Sometimes it helps to slide the blades out, so that you are only paddling with the blade of one paddle on each side. People with small hands will find this difficult. If your boat has a tow system, you can slide the retrieved paddle under the line on the rear deck and keep one blade "chicken-winged" under an armpit.

Boats The "bulldozer" technique is the most common method of boat recovery, as it requires little in the way of special equipment or training. The paddler in the recovery boat places his bow on the downstream side of the abandoned boat, sets up as if for a ferry, and then pushes the boat into a convenient eddy. Paddlers recovering decked boats should try to put their bow inside the cockpit of the swamped boat, making sure they will be able to withdraw if the need arises. Open canoes and rafts sometimes have a problem keeping contact with a partly submerged boat.

A decked boat can often be flipped back upright before it fills completely with water, which makes it much easier to handle: if there is not much water in it a good shove can send it into a nearby eddy. Open canoes can also be recovered this way, and unless the boat is completely swamped it is definitely advantageous to try to turn it upright.

Fig. 5.5. Most paddlers use the "bulldozer" method to recover boats. It's convenient but has its limitations. A half-submerged boat, like this one, can be difficult to push. Tow systems work better.

A safer and better way to recover a boat is to tow it in with a towing system or painter (see Chapter 2), because the recovering boat can stay upstream and avoid being caught between the swamped boat and a rock. Tow systems are not yet widely used by recreational paddlers. You do see them on the decks of instructors' boats and on expedition boats — these paddlers cannot afford to lose a boat on some remote river. However, tow systems have many applications other than boat rescue: they are easier to hang on to than grab loops if you are swimming; they can be used to haul a boat up an embankment; and they make it very easy to tie off a boat instead of pulling it up on shore. As the word spreads, tow systems will undoubtedly become more popular.

Fig. 5.6. The painter on the tow boat is run through the other boat's grab loop and then secured with the jam cleat . . .

Fig. 5.7. . . . and then the boat is towed away. All tow systems should have a quick release.

Boats can also be used to rescue other boaters from hydraulics. For small hydraulics the paddler in the rescue boat should stick his bow into the backwash so that the "stuck" paddler can grab it (Figs. 5.11-5.12). In larger hydraulics the rescuer may try approaching from upstream with maximum momentum and drop in sideways onto the stuck boat, knocking it clear of the backwash. This technique presupposes a willingness on the part of the rescue boater to enter the hydraulic the stuck boat has just left. Neither technique is for the faint of heart.

Rescuing Swimmers

People who think their life is in danger often act irrationally. We have seen a panicked swimmer crawl onto a rescuing kayak, tip it over, and then crawl up onto the upside-down boat, preventing the kayaker from rolling and providing the river with an additional victim. Approach any swimmer with caution.

Decked Boats   If time permits, the best thing to do is approach the victim so that you are just out of his reach and try to get his attention. This way you will be able to make some assessment of his mental and physical condition and give him some instructions. Do *not* paddle up beside the swimmer, because he may grab you or your paddle and tip you over. Appear calm and in control, but be on guard for the unexpected. Offer the swimmer the bow or stern grab loop of your boat. If you're going to tow him any distance, a stern hold makes this easier. Tell the swimmer not to climb up on the boat (there are exceptions to this, which we will discuss later) and to help by kicking with his feet.

Rescue in a major rapid or in big water calls for some extra precautions. The violent up-and-down movement of the boat can easily smash a swimmer's face, even if he is hanging on to the grab loop. In a drop-and-pool river it is often better to follow the swimmer and offer assistance at the bottom. Having someone nearby in the rapid can be very heartening. In big, continuous water this is more difficult. Western boaters often use a third grab loop or a "Blackadar handle" of wood near the cockpit so that the swimmer can climb onto the rear deck. This avoids problems with the stern hitting the swimmer's face, makes it easier for him to

Fig. 5.8. The basic method of rescuing a swimmer with a kayak is to have the swimmer hang on to the grab loop and kick vigorously with his feet.

breathe in big waves, and somewhat reduces his exposure to cold water. A less satisfactory method, because of the danger of popping the sprayskirt, is for the swimmer to hold onto the cockpit rim. It may be best to put D-shaped handles (screen-door handles are good; see Fig. 5.13) on both sides of the deck just aft of the cockpit to afford a balanced grip.

Having an extra person on the back deck makes a boat heavy and unstable, but the victim can assist by dropping his legs in the water to act as "daggerboards." Before starting, tell the victim to let go of the boat if it tips over, so you can roll.

Open Canoes Approach victims with even more caution in an open canoe than in a decked boat — for most canoeists there is no possibility of an Eskimo roll. In easier water you can rescue a swimmer from an open boat by throwing out the stern painter for him to grab. If you have two paddlers, this will increase both your speed and towing power.

The open canoe can be very useful as the basis of the Telfer lower rescue system described later in this chapter. It also does

Fig. 5.9. Having the swimmer climb up on the back deck may be necessary in bigger water. Here the swimmer grasps the cockpit rim (a handle would be better) and holds his feet up to reduce drag.

Fig. 5.10. If this makes the boat too tippy the swimmer can drop his legs to act as "daggerboards," although this presents more resistance to the paddler. Grabbing the cockpit rim may cause the spray skirt to pop off.

Fig. 5.11. *To rescue another paddler from a small hole, this paddler has given his friend the bow of his boat. By backstroking . . .*

Fig. 5.12. *. . . he gets the paddler but not the boat. Results with this method tend to be unpredictable.*

some of the more mundane tasks of rescue quite well: ferrying ropes, equipment, and people across calm stretches, or acting as a stretcher or shelter in a pinch.

Rafts Rafts have the potential to produce swimmers in large numbers, and they are particularly slow and hard to manage after they have filled with water and lost one or more crew members. This makes it particularly important to train raft crews in rescue techniques.

The best time to rescue a person who has fallen out of a raft is within the next few seconds. Even in very turbulent water the swimmer is usually still near enough to the raft to be reached with a hand, an extended paddle, or an oar. If the raft is in a rapid, only those crew members who are directly involved in the rescue should stop paddling, unless it becomes obvious that more help is needed. Some rafters, especially in oar boats, are beginning to use a short (15- to 25-foot) throw bag for rescue. This is clipped into a convenient D-ring or worn on the boatman's waist; a raft captain can make a quick toss and then go back to rowing immediately. The swimmer can pull himself in or be dragged in by other crew members. A paddle-raft variation is to have a "baby bag" secured within easy reach of the raft captain. All such systems need to be attached with a quick-release clip.

The first thought of a raft crew, assuming the raft itself is not in danger, should be to get the swimmer *back in the raft:* clinging to

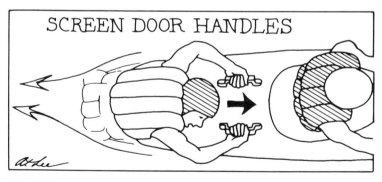

Fig. 5.13. Rescue of a swimmer by boat.

Fig. 5.14. Pull a swimmer into a raft by grabbing the shoulder straps of his lifejacket and falling back into the raft, not (as the people on the left are doing) by pulling on his arm.

the side of a raft feels secure, but it can be very dangerous. It is not unusual to see people laughing and talking as they float down the river like this, oblivious of the danger of being crushed between the raft and a rock. The best way to pull a swimmer into a raft is to stand up, brace your knees against the tube, and grab the shoulder-straps of the lifejacket. Then lean back and pull the swimmer in. If you grab the swimmer's hands or arms you will only drape his body over the tube. The swimmer can assist by doing a scissors kick and pulling up on a convenient D-ring while being lifted.

If the raft flips on a big river it may be possible for some crew members to climb on top, rescue the other members, and then paddle the upside-down raft to the shore.

The Telfer Lower

The Telfer lower is a mixed system of rescue: it uses a boat or boats but is controlled partly from the shore. A working platform made with a raft or a combination of boats is attached to an anchor line stretched across the river and is floated downstream to the accident site. Use the Telfer lower when more conventional means of rescue (rope throws, ferries, and tag lines) are not feasible or have failed. Even in very swift water it provides a surprisingly stable platform from which to rescue a trapped victim or a pinned boat. The disadvantages of this rescue method are that it takes time to set up and taxes the technical skills and leadership of the rescuers. It also requires a substantial amount of equipment, particularly rope.

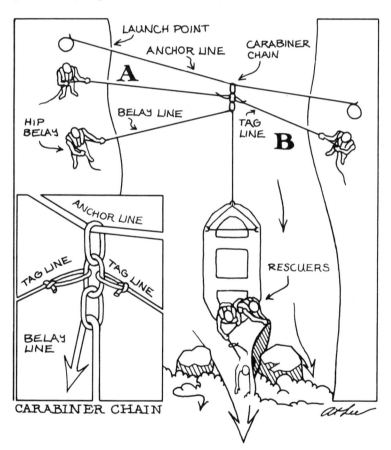

Fig. 5.15. The Telfer lower: Belayer on shore. Tag line A may be eliminated if the rescuers are shorthanded.

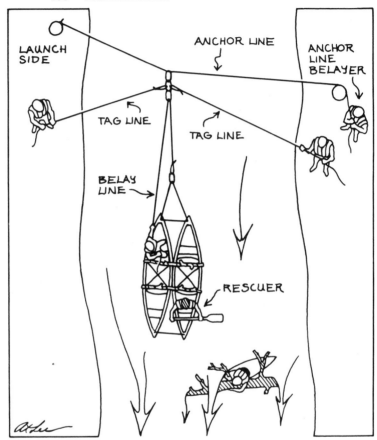

Fig. 5.16. The Telfer lower: Belayer in rescue craft, using the bottom of the carabiner chain as a pulley for the belay line. The belayer may hold the line or use the thwart as a belay point. The anchor-line belayer may be eliminated if the rescuers are shorthanded.

Setup   Establish an anchor line across the river 20 to 50 feet upstream of the accident site. You can throw it across a narrow river or ferry it across a wide one. The anchor line should be 8 to 10 feet above the surface of the water and as tight as possible. Trees usually make the best anchor points, but you could use boulders, car bumpers, or other fixed objects instead. The anchor line should be angled downstream, so that the current will carry the rescue craft toward the accident site.

The rescue craft can be constructed in a variety of ways. A raft works best for heavy water, as long as it is fully inflated and rigged with stout D-rings. Two open canoes can be lashed together catamaran-style about a foot apart, using saplings or paddles to keep them together. They will resist the current less than a raft but are more likely to swamp. A canoe and a kayak can be used, as can two kayaks. A single boat will also work, but it will not be as stable as a raft or two boats.

The heart of the Telfer lower is the carabiner chain (Fig. 5.15). Using three carabiners provides a flexible joint that will slide

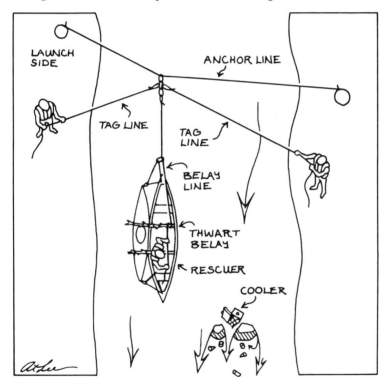

Fig. 5.17. The Telfer lower: The rescue craft is lowered or raised with a single line attached to the lower carabiner.

easily on the anchor line. The upper carabiner functions as a pulley on the anchor line, the lower carabiner serves as the belay pulley for lowering the rescue craft to the accident site, and the middle carabiner links the two and provides a tie-in point for tag lines. The tag lines on the middle carabiner control the left-to-right position of the rescue craft in the current. If the anchor line is slanted downstream you can use a single tag line on the launch side, but lines on both sides will give more precise control.

Lowering There are three ways to lower the rescue craft to the victim:

- *Belayer on shore.* The rope is attached to the boat and run through the lower carabiner to the shore party, who belay according to instructions from the boat. This method allows the rescuers in the craft to concentrate their efforts on the victim or boat but requires a substantial amount of rope — you cannot use two ropes joined together, because a knot will not pass through a standard carabiner. There may also be communications problems between the shore party and the rescue craft.

- *Belayer in rescue craft.* The lowering rope is attached to the rescue craft, runs through the lower carabiner, and then runs back to the rescue craft. This method gives the rescuers precise control and eliminates the communications problems in the first method. It requires less rope than the first method but more than the third. The pulley system gives an effective mechanical advantage of about two to one.

- *Belayer in rescue craft.* The rope is tied to the lower carabiner and the rescue craft is belayed directly by the rescuer. Of the three methods, this requires the least rope but is the most difficult, since there is no mechanical advantage. Belay by wrapping the rope around a D-ring, around a thwart, or through a carabiner (using a Munter hitch). Avoid using a hip belay.

The choice of method will depend on the number of people, the

Fig. 5.18. The Telfer lower ready to go. The rescue craft is made with two canoes lashed together. Note that the anchor line is angled downstream so that the current pushes the boats toward the rescue site.

Fig. 5.19. This rescue craft is lashed together with paddles. Longer paddles (like kayak paddles) would be better. This craft is set up for a direct lower as shown in Fig. 5.17. Note the V-harness on the bow to keep the craft straight.

Fig. 5.20. The whitewater sleigh ride! This version of the Telfer uses the belayer in the rescue craft. The rope is run from the carabiner on the V-harness in the bow to the lowering carabiner and then back to the belayer in the boat.

Fig. 5.21. The direct lower with the belayer in the rescue craft. The belayer can either hold the rope directly, as here, or use a friction belay.

number and type of boats, and the amount of rope available, as well as the distance the rescue craft must be lowered.

The Rescue   The crew of the rescue craft should have paddles to control the angle of the craft and to paddle it out if the system fails. All rescuers should have lifejackets and knives. Use pre-arranged hand signals to coordinate the movements of the boat. There should be a team leader in the boat, but the rescue leader should remain on shore if possible to oversee the entire operation. Someone should stand upstream of the rescue site to warn off other boaters, and there should also be rope throwers downstream. Put the best paddlers in the boat.

   Make sure the bow of the boat remains pointed directly upstream, because if the boat swings broadside to the current it is

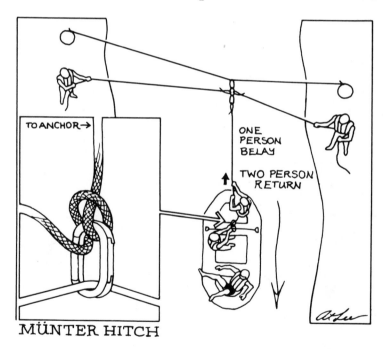

Fig. 5.22. The Telfer lower, using a Munter Hitch for positioning the rescue craft upstream or downstream.

much more likely to get swamped. If the boat does get swamped, the belayer and the tag-line team should try to pendulum the boat ashore. If this is not possible the rescue craft must be released, and there must be no knots or bags on the line to impede the release if this is necessary. The rescue crew should not hesitate to cut the rope.

It is dangerous to the rescuers to be on the upstream side of the hazard. If possible, the Telfer lower should be rigged on the downstream side of the obstacle and the floating platform should be raised to the accident site. This can be done with a shore-based rope team or by using a Munter Hitch (Fig. 5.22). One person hauls on the upstream side of the hitch while the other takes up the slack with the Munter Hitch. It is slow and difficult work.

Conclusion

Under the right conditions, boat-based rescue is very effective; in some situations (very wide rivers, for example) it will be the primary rescue system. Boat-based rescue can also be effectively integrated with shore-based rope systems. The Telfer lower, however, uses elements of both systems and can be used effectively both as a rescue platform and as a means of ferrying patients across rivers during evacuations.

Wherein I spoke of most disasterous chances,
Of moving accidents by flood and field,
Of hair-breadth 'scapes i' the imminent deadly breach.

WILLIAM SHAKESPEARE, *OTHELLO*

· 6 ·
Entrapments and Extrications

A group of expert paddlers went out on North Carolina's Watauga River in the spring. The Watauga is steep, with almost continuous rapids, many big drops, and a 15-foot waterfall. "You had to catch every eddy in order to boat scout the next rapid," one member of the group recalls, "and we didn't know exactly where the waterfall was." As it turned out, the eddy that one of the paddlers missed was the one just above the waterfall. He went down backwards over the wrong chute and pinned halfway down the falls on a small boulder.

Fortunately, he was with an alert, experienced group of paddlers. They immediately sized up the situation, got out of their boats, and went in after him. The entrapped paddler was able to breathe, but the water was very cold. Though he was dressed for the season, hypothermia was an immediate danger. His legs were entrapped under the thwart seat in his low-volume C-1. The rescuers were able to wade out to the boulder, at some danger to themselves, and hold the victim's head out of the water. "Finally," said one of the rescuers, "we were able to pull him up enough to let him slide his legs out from under the thwart and get free. But it was touchy and he wouldn't have lasted long in that cold water."

Here we have the classic elements of both a boat pin and an entrapment: a good paddler in hard water making only a small mistake. In this case there was a happy ending, but it was only because of the quick and correct actions of the rescuers.

Entrapments and Boat Pins

People are entrapped; boats are pinned. "Entrapped" means being held in a life-threatening position by the force of the water or by a collapsed boat; a boat is "pinned" when it is held in place against a solid object by the force of the current. Since an entrapment is often the direct result of a pin, we will discuss them together here. (Foot entrapment is discussed along with rescue methods in Chapter 4.)

Boat pins fall into two main categories: broaches and vertical pins.

Broaches A boat broaches when it wraps sideways around an obstacle, usually a bridge piling, a boulder, or a tree. There is usually a cushion of water on the upstream side of an obstacle, and the river-wise paddler can normally ride over this "pillow" and avoid a direct collision. Some obstacles, however, like bridge

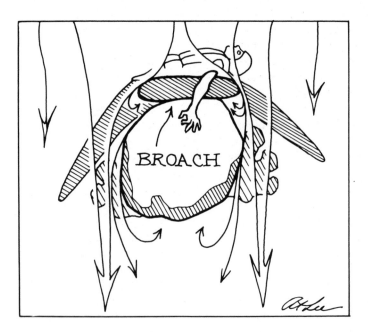

Fig. 6.1. A broached boat.

pilings and undercut rocks, have little or no upstream cushion and are much more dangerous to approach.

The fatal moment of the broach occurs when the boat collapses and entraps the paddler. This most often happens to kayakers—the front deck of their boat collapses on their legs — but rafts and canoes can entrap people as well.

Vertical Pins   A boat is vertically pinned when the bow slams into the riverbed and stays there. This is most common on steep, ledgy, shallow rivers. Paddling drops of more than 4 or 5 feet invites a vertical pin, but boat design also plays a part. A low-volume needle-nosed racer is more prone to pinning vertically than a blunt-bowed cruiser. People who habitually paddle very steep creeks often use large-volume designs that would be considered obsolete elsewhere.

When the bow is wedged into rocks or under a submerged tree, the force of the water will quickly produce catastrophic results: the boat may "pitchpole" and collapse at the fulcrum created by the obstacle, but a more common result is for the water to force the stern of the boat to the bottom too, collapsing it along with the buried bow. Even if the boat does not actually fold, the paddler may be held inside the boat by the sheer force of the current against his back. If the water is not too deep he may be

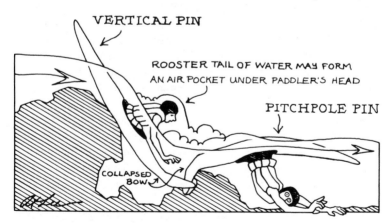

Fig. 6.2. Vertical and pitchpole pins.

able to form an air pocket in front of his body by pushing up on the front deck. This is very tiring, especially in cold water, and rescue efforts must begin immediately.

C-1s (decked canoes) are often cited as being less likely to entrap a paddler than a kayak. Since the paddler's legs are tucked under him, so the argument goes, he has less exposure than a kayaker whose legs are fully extended. With today's low-volume canoe designs, however, this is no longer entirely true. In a small boat, particularly one with a thwart rather than a pedestal seat, a canoeist's feet can be entrapped by the collapse of the *rear* deck. Open canoes are less vulnerable to entrapments, but they are certainly not immune. One open-boat paddler spent over two hours under Sweet's Falls on the Gauley River after pinning in the drop: the boat collapsed, pinning one of his legs between the crushed hull and a thwart; another canoeist nearly drowned on the Ocoee River after broaching his canoe on a bridge piling. He became entangled in his thighstraps and had to be cut out by a rescue party (see Chapter 7).

Entrapment Rescues

An entrapment rescue generally proceeds in four stages: assessment, stabilization, extrication, and evacuation.

Assessment   This is a critical first step. The most important question is whether it is a "head-up" or "head-down" entrapment: if the victim can breathe you can afford a few more minutes to decide on and use the best method of rescue; if the victim's head is underwater you must react immediately. (Timing and organization are discussed in more detail in Chapter 8.)

Stabilization   Both the condition of the victim (who becomes a patient once you reach him) and the position of the boat must be stabilized. Don't charge blindly into trying to remove the patient before you have made sure both are stable. Make sure first that the patient can continue to breathe; then see that the boat does not shift and make the situation worse.

If the entrapment site is accessible, the quickest and most obvious way to stabilize the situation is with a "hands-on" method

Fig. 6.3. Rigging a line to keep the victim's head out of the water. (AWA photo)

like the one described at the beginning of this chapter. Consider whether the rescuers can stand on or at least near the obstacle that created the entrapment. If they cannot stand in the current, can a rescuer be pendulumed down to the site by rope and effect a strong-swimmer rescue (as discussed in Chapter 4) or hold the patient's head above water? Can a rope from shore or a tag line be rigged so that the victim's head will stay above water? And should auxiliary lines be tied to the pinned boat to keep it from shifting?

If the rescuers cannot get close to the victim, another possibility is to use a floating tag line (see Chapter 4). Bring the tag line up from the downstream side, so that the victim can drape his arms over it, then pull it upstream to help keep his head up. Even if the victim cannot hear or see under a rooster tail of water, the tag line will provide the psychological boost of letting him know someone is out there trying to help.

Extrication The extrication process should not start until the victim has been stabilized in a relatively safe condition. If he cannot be stabilized, however, as in a head-down pin, you must begin extrication efforts immediately.

ROOSTER TAIL

AIR POCKET

Fig. 6.4. A stabilization tag line can be set up to keep the pinned paddler's head up while other rescue methods are used to extricate him.

Consider first whether the boat can be cut to release the victim. Modern boat materials make this difficult, but especially with a raft it may be the logical solution. Most extrications, however, will involve attaching lines to the pinned craft and then using shore-based power to pull the boat free. (Methods of extricating pinned boats are discussed in more detail in the Boat Recovery section below, and the various rope tricks discussed in Chapter 4 can also be used to stabilize or extricate a victim.)

Situations that involve foot entrapment are similarly handled. The normal way to extricate a victim whose foot is entrapped is to use a floating tag line or a snag tag (see Chapter 4).

Evacuation The final stage of an entrapment rescue is the evacuation of the patient from the entrapment site to the shore. Monitor the patient's condition continuously from the time the rescuers reach the scene and continue to do so throughout the evacuation. Use CPR and first aid as necessary.

You should give some thought to the evacuation of the patient while you are involved in extricating him. If the patient is coherent and uninjured, you might simply swing him in with a rope; if he is injured but has been rescued from immediate danger, you should use other, slower methods for evacuation. You might pick him up with a raft or another boat or by means of a Telfer lower. If the situation requires it and conditions are suitable, he might even be evacuated by helicopter directly from the entrapment site (see Chapter 9).

Recovery of Pinned Boats

Extricating a pinned boat can be an interesting project — so long as there is no victim involved and the safety of the group does not depend on success. Usually the damage is already done and there is little need for haste. Opinions can be exchanged, options weighed, and the time taken to make sure everyone

Fig. 6.5. The "armstrong" method will work in a lot of cases. Just grab the boat and start pulling. Give some thought to what will happen after the boat comes free.

understands the retrieval plan. Since time is not critical, though, use the safest means of extrication, bearing in mind not only your own group but also other paddlers on the river. Try the simplest methods first; go on to more complex systems if the simpler ones fail. Often it is a combination of techniques that works best.

"Armstrong" Method   The quickest and simplest method is just to grab the boat and start pulling. This works best if you can safely stand on the river bottom or on the obstacle causing the pin. A strong back, or several, helps. Often lifting one end of the boat while jumping up and down on the other end will work. Most people's legs are stronger than their arms, and it may be more effective to sit on the obstacle and push with your legs. If you do this, though, don't let yourself slip down between the boat and the obstacle.

Sometimes a simple lever, such as a log or tree limb, can be used to pry the boat off the obstacle. A rope can be tied to the high end and run to shore for additional people to pull on. Tie a backup rope to the boat to swing it into shore when it comes loose.

Rigging

Often the hardest part of a boat recovery is getting the rope attached to the boat. If the current was strong enough to pin the boat, it is probably too strong to stand in. Working around pinned boats in swift water is risky and demands caution: the boat may shift at any time, but it is especially prone to do so under tension from the haul line.

A pinned boat will usually be near the surface, and you can often stand on the boat itself or the obstacle that caused the pin. If you cannot, you will have to try another method of reaching the pinned craft, such as a Telfer lower or a Tyrolean traverse. A completely submerged boat, on the other hand, presents a real problem. Sometimes you can use branches or paddles to push the rope under the boat and then retrieve the loose end on the downstream side of the boat. You can also drop lines weighted with throwbags or ammo boxes stuffed with rocks above the boat and pull them out below with the T-grip of a paddle. The rope tricks mentioned in Chapter 4 should also be considered.

Fig. 6.6. Sometimes you'll be lucky enough to have a big flat rock to work on; sometimes you can even tie in on the canoe's thwarts.

Fig. 6.7. Other times there may not be any thwarts, and you'll have to make a cradle rig.

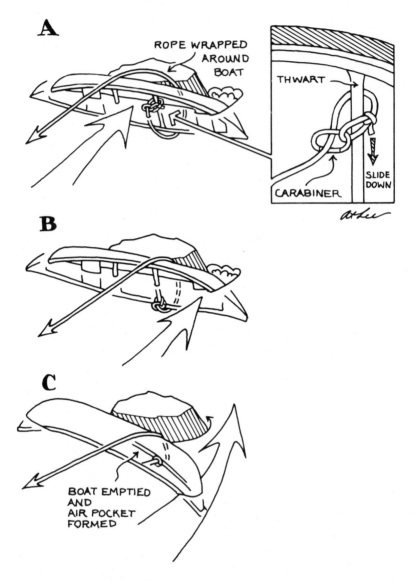

A

ROPE WRAPPED
AROUND
BOAT

THWART

CARABINER

SLIDE
DOWN

B

C

BOAT EMPTIED
AND
AIR POCKET
FORMED

Fig. 6.8. A carabiner clipped into the end of a line to form a loop is a quick way to attach the line to a thwart.

Thwarts on canoes and grab loops on decked boats are the obvious points of attachment. On rafts a thwart is usually a stronger attachment point than a D-ring. Nevertheless, it is sometimes better, if the boat is severely pinned, to encircle the boat with the line rather than attach it to a single point. If the thwarts, grab loops, or D-rings are missing or inaccessible, this is your only choice. One quick method of encircling the boat or attaching a line to a raft thwart is simply to tie a carabiner to the end of the line, pull it around the boat, and clip it into the line (Fig. 6.8). You can also use a cradle rig: two loops of rope, each of which is smaller in diameter than the widest point of the boat. Slip these over both ends of the boat and connect them with another piece

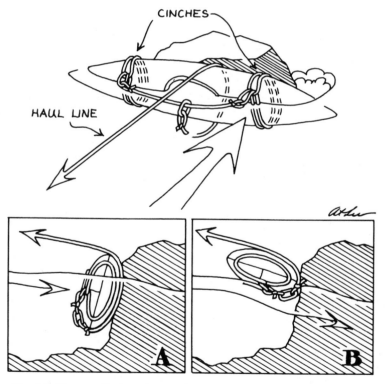

Fig. 6.9. Use a cradle rig on boats without thwarts.

of rope. Attach the haul line to the connecting piece (Fig. 6.9).

Open canoes and rafts typically pin with the open side up-stream and the bottom against the obstacle. If possible, roll the boat over as you pull it off the obstacle, so as to dump the water and reduce the "sea anchor" effect. One way to do this is to attach

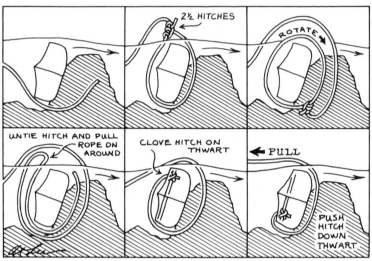

Fig. 6.10. The Steve Thomas rope trick.

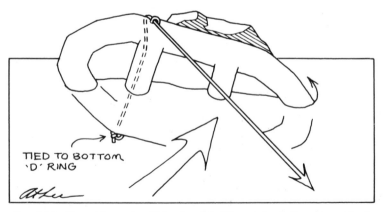

Fig. 6.11. Pinned boats should be rigged to flip and dump the water as they're rolled out. (See also Figs. 6.8-6.9.)

the rope at the low end of a thwart, pass the rope between the hull and the rock, and pull so that the water spills out. With rafts, attach the haul line to the D-ring closest to the river bottom but on the far side of the boat (Fig. 6.11). As you pull on the rope it will apply a rotating force (torque) on the raft and cause it to rotate on its axis as it dumps the water out.

If you find you cannot dump the water out of a raft, consider deflating one or more of the compartments. Air valves can be bled underwater without getting water in the tubes, as long as the air pressure is greater than the water pressure. If you have a good stream of bubbles the air pressure is adequate. It is a good idea to unload the raft before taking this drastic step.

As a last resort, cut through the floor of the raft to reduce the force of the current. This may free the raft, but it will mean repair time later. Most pinned rafts can be freed without cutting the floor.

The Force of the Current

When rigging for a boat recovery, many people forget to consider the force and direction of the current. Time and again we have seen elaborate haul systems based on convenient anchor points or work areas rather than angle of pull. To some extent, of course, the direction of any pull is dictated by where you can anchor it, but the idea is to make the current work for you. Keep trying: sometimes only a small change in the angle of pull will make the difference.

Do not blindly trust mechanical devices, no matter how powerful they are. A common mistake is to assume that the problem is simply not enough pull. An electric winch or a well-set-up Z-drag can literally pull a boat to pieces.

If the situation permits, several ropes pulling from different directions are usually more effective than a single rope. Multiple rigging can be used to avoid pulling directly against the current. For example, one rope can be used to relieve the pressure on the boat and another to slide the boat sideways until it clears the obstacle. Still another line might be used to flip the boat so as to dump water out of it, as described previously (Fig. 6.13).

When possible, combine hauling systems. The vector pull, for

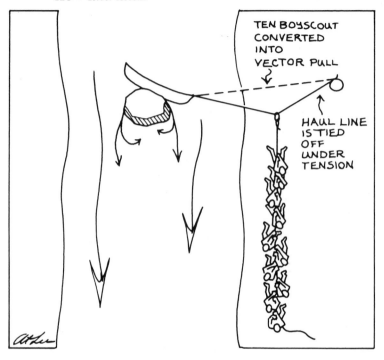

TEN BOYSCOUT CONVERTED INTO VECTOR PULL

HAUL LINE IS TIED OFF UNDER TENSION

Fig. 6.12. A vector pull will provide some mechanical advantage with a slight change of direction.

example, can be a useful addition to any of the haul systems described below. It will give some mechanical advantage and change the angle of pull, too.

To change the direction of the pull, you might use a directional pulley. This allows the hauling team to pull in the most convenient direction, even if the accident site would not otherwise allow it (Fig. 6.14). One imaginative group of rescuers put the pulley on a tree limb *above* the pinned canoe, which enabled them to use their body weight to pull directly down on the haul line.

Haul Systems

Most systems of recovery other than the "armstrong" method use a rope or line attached to the boat. These are collectively called haul systems. The systems described here are meant to get the boat off the obstacle. Remember to consider what will happen once the boat comes free, since water-filled boats, particularly rafts, are very heavy. Backup belay lines to the craft may be needed *in addition to* the haul lines.

"Ten Boy Scouts" Method This is a simple rope pull with no mechanical advantage: attach the rope to the boat, find the best angle to pull from, and pull. If the boat doesn't move, just add more boy scouts. This is the most common method of extrication and is easily integrated with other systems. The force of a direct pull can be increased to the breaking strength of the rope by using a mechanical device like a "come-along" or winch. Most recreational paddlers don't carry these, but they may be available if an outfitter or search-and-rescue unit is on the scene.

Vector Pull Pushing or pulling in the middle of an already taut line will add some mechanical advantage, as well as changing the direction of pull slightly. With your ten boy scouts, haul as hard as you can on the line and then tie it off under tension. Then tie or clip a second line into the center of the haul line and pull again at a 90-degree angle to the first. This is effective until the haul line is at an angle of about 30 degrees to its original position. The vector pull can easily be used with other systems and is usually the next thing to try if a direct pull fails.

The Z-drag The haul system most familiar to river veterans is the Z-drag, a pulley system that may be rigged for several degrees of mechanical advantage. Borrowed from rock climbing, the Z-drag is quite flexible and can be set up quickly, with a minimum of equipment. The simplest Z-drag gives a theoretical mechanical advantage of 3:1; more complex Z-drags give a theoretical mechanical advantage of up to 27:1. The Z-drag is quick to set up and can be integrated with other systems.

The pulleys referred to in this section will usually be carabiners. A real pulley is ideal for rescue work, because of its low friction, but carabiners can be and usually are substituted. Carabiners add friction and therefore reduce the efficiency of any pulley system, but they are much more easily carried than pulleys. If no carabiners are available a rope loop made with a butterfly or figure-of-

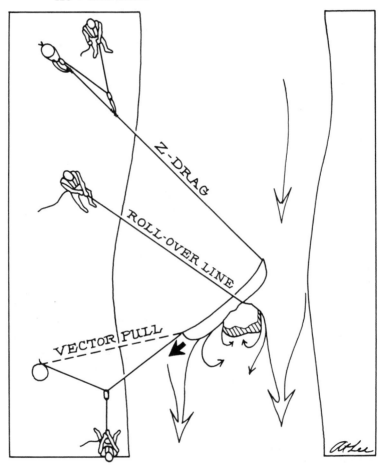

Fig. 6.13. Several haul lines are usually more effective than one. Here a Z-drag is used to pull the boat away from the rock, while another line is used to roll the boat and dump the water. A third line slides the boat along the face of the rock and into an eddy.

eight knot can serve as a pulley, but the friction is *much* greater and there is a real danger that the haul line will saw through it. (For the sake of comparison, the actual mechanical advantage provided by a theoretical 3:1 pulley system would be between 2.5:1 and 2.7:1 with real pulleys, 2:1 with carabiners, and even lower with rope loops.)

After attaching the haul line to the pinned boat and figuring the angle of pull, locate an anchor point. A tree or boulder works well but any secure anchor, such as a car bumper, a root, or a log can be used instead. Remember, though, that the anchor point has to hold the whole load. Tie an anchor sling (a strong piece of rope will do) around the anchor point to hold the primary pulley. Run the haul line through the primary pulley at the anchor point, then through a second pulley (the "traveling pulley"), and finally back to the shore. Attach the traveling pulley to the haul line between the primary pulley and the pinned boat with a butterfly or figure-of-eight knot or with a prusik.

The polypropylene ropes used in river work are very elastic, so there will be a lot of stretch before the boat moves, and since the traveling pulley moves back toward the anchor point as the rescuers haul on the line it must be periodically reset further down the haul line while still maintaining tension on the boat. You can do this with a separate belay line or with a prusik, of which the prusik is the more effective. Usually made with small-diameter (5- to 7-mm) nylon rope with the ends tied together in a double fisherman's knot to form a loop (see Appendix D), the prusik is useful because it will slide freely on the haul line when there is no tension but will grip when it has a load on it.

A typical Z-drag system uses two prusiks: one to secure the traveling pulley to the haul line (the traveling prusik) and the other as a brake (the brake prusik). With a prusik on the haul line

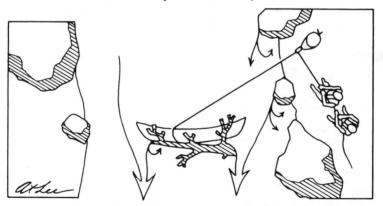

Fig. 6.14. A directional pulley will change the direction of pull but offers no mechanical advantage.

Fig. 6.16. A prusik knot holds the traveling pulley to the haul line. The prusik will grip under tension but can be slid down the line to reset the pulley when the tension on it is released.

Fig. 6.15. This Z-drag is pulling against the boat pictured in Fig. 6.6. It's been set to pull at an angle to the current.

you can easily slide the traveling pulley back down the line to reset it and not have to worry about trying to slide a tied rope loop through the anchor pulley.

Before sliding the loop for the traveling pulley down the haul line, you will have to release the tension on it. Rig the brake prusik to grip in the direction opposite to the braking effect of the traveling-pulley prusik. When the pull on the haul line is relaxed, as when you are resetting the traveling pulley, the brake prusik will grip the haul line and maintain tension on the system. The disadvantage of this system is that the brake prusik may be hard to release when under tension, and there are times when the system is under so much strain that the only thing to do is release the boat. This may mean cutting the prusik. An alternative to the brake

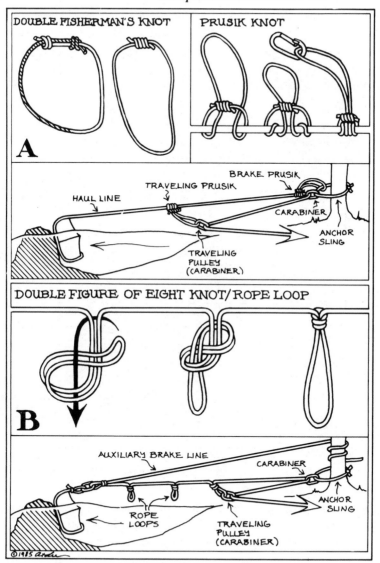

Fig. 6.17. *Two types of simple Z-drag. Type A uses prusik knots to slide down the haul line, while type B uses a series of rope loops. Type B requires a separate brake line to hold the load while the line is reset to the next loop.*

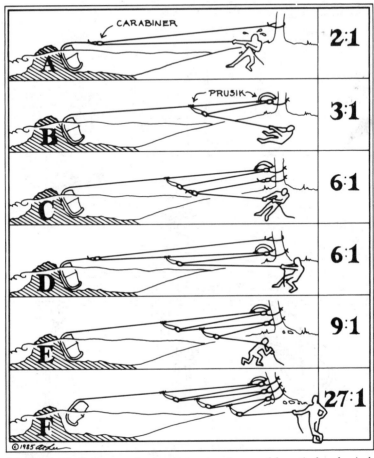

CARABINER

PRUSIK

2:1

3:1

6:1

6:1

9:1

27:1

©1985 a. L.

Fig. 6.18. The Z-drag may be rigged in various degrees of theoretical mechanical advantage from 3:1 to 27:1.

prusik is a separate belay line to maintain tension on the pinned craft while you are resetting the traveling pulley. Both the brake prusik and the backup belay rope require someone to monitor them.

To multiply the mechanical advantage of the Z-drag system, increase the number of pulleys: a double Z-drag gives a 9:1 and a triple Z-drag a 27:1 theoretical mechanical advantage. Only one brake prusik, attached to the haul line from the boat, is necessary for any of these systems. (For the sake of clarity we have shown each anchor pulley as separate, but in fact a single pulley can take two lines.)

Although these rigs will increase the pulling force on the haul line, there are some problems. Friction increases as you increase the number of carabiners, and it becomes difficult to stop things getting tangled up when you use complicated systems. If the mechanical advantage is great, a long pull on one end of the system means a short pull on the other end, and the several pulleys must constantly be reset. Remember: it is sometimes better to try a different angle of pull than to increase the pulling power.

The Z-drag seems complicated at first, but with practice it can be set up and used quickly, and it has other practical applications, like pulling your stuck shuttle car out of the mud on the take-out road.

Fig. 6.19. The brake prusik, located at the anchor pulley, holds the haul line when tension on the system is released to allow the traveling pulley to be reset. Someone must slide the brake prusik down the line periodically during hauling or it will bind.

The Sea-Anchor Haul System   This is an ingenious haul system suggested to us by an Alaska wilderness guide. It takes advantage of the force of the current against an open canoe to provide the pulling force to get the pinned boat off an obstacle. Attach a haul line to the pinned craft, then rig a directional pulley upstream of it to give the correct angle of pull. Then run the line to where the sea anchor will be set up. The sea anchor itself is usually an open canoe, but with a little ingenuity other kinds of craft could be used. Attach the canoe to the haul line with a cradle rig so as to be sure the hull and not the thwarts will take the force. The rescuers will have to maneuver the overturned canoe into the current to load the haul line. If the current on the sea anchor is equal to or greater than the current holding the pinned boat, the boat should come off the obstacle. In particularly stubborn cases it might be possible to use a second sea-anchor canoe to add pulling power.

Both the pinned canoe and the sea anchor should have backup belay ropes attached to them to ensure their timely return. Since there is some danger of losing or pinning the canoe serving as the sea anchor, try other methods first.

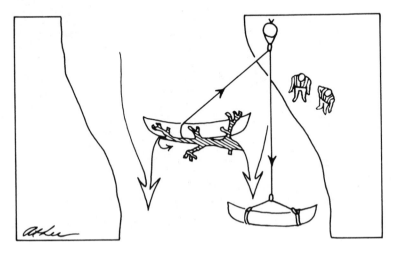

Fig. 6.20. The sea-anchor haul system.

Conclusion

Les describes an incident in which a little thought about the direction of pull made all the difference. "We had a major problem. One of our rafts was pinned on a rock in the center of the Ocoee River at Broken Nose. All our people were on the right shore. It was the wrong angle of pull, but we figured with all our manpower we ought to be able to muscle it off. We were wrong: a double Z-drag and fourteen people couldn't budge it, and our haul lines were creating a serious hazard for other rafters coming down the river. We took the Z-drag down and sent the rest of the trip on while another guide and I stayed to deal with the raft.

"There were some boulders directly upstream of the pinned raft. A Z-drag rigged from here would still not be at the ideal angle, but it would be better than what we had previously tried. The boulders didn't have any trees and I couldn't rig a sling because of the lack of rope. One of the boulders did have a crack and I was able to jam a knot into it to create an attachment point for the anchor pulley. We rigged a simple Z-drag, but still the raft wouldn't move. Finally, with the haul line still under tension, we tried a vector pull by just pushing in on it with our hands.

"Imagine our surprise when that raft came off! Especially with just the two of us working on it after fourteen had failed. It showed us what the right angle of pull and some trial and error could do."

Fig. 6.21. Pinned! And solidly, too. Although the guide was unable to prevent the pin, he has gotten his crew safely out of the river.

Fig. 6.22. The rescue crew moves in and begins to set up a Z-drag on some rocks in the center of the river. The haul line is attached to one of the submerged raft's thwarts.

Fig. 6.23. Setting the anchor point on a very strong little tree, they haul away . . .

Fig. 6.24. . . . until the raft comes off.

Fig. 6.25. Another Z-drag, this time on the Chattooga. The Z-drag is set to pull at an angle to the current and is attached to the far thwart of the raft. This will tend to slide the raft along the face of the rock. A worn-through pothole is used as an anchor point. (Photo by Mary Ellen Hammond/Nantahala Outdoor Center)

Fig. 6.26. When the raft comes off, it is belayed by wrapping the haul line around the anchor point. No brake prusik was used in this setup. (Photo by Mary Ellen Hammond/Nantahala Outdoor Center)

Oh to be rescued
from above
On the wings
of a dove
TRADITIONAL

· 7 ·
Vertical Rescue

A vertical rescue involves lowering the rescuer to the accident site. It is usually a technique of last resort, because of the time required to organize equipment and manpower and because, like the strong-swimmer rescues, it puts the rescuer in a potentially dangerous position. In fact, a vertical rope rescue has definite setup limitations. Tyrolean rescues, for example, involve stretching a line across the river (which becomes difficult if the river is over 200 feet wide or if the rope ferry must be done in heavy water) and require anchor points such as trees or boulders at the accident site. Other vertical rescues, such as bridge lowers, require that the vertical structure you are going to use as a base be at or very near the accident site.

Yet sometimes there is no other way to get to the victim. If the water is too heavy for a Telfer lower or a boat rescue, or if there are no convenient eddies, it may be the only available technique. A low-head dam may have a catwalk above it from which a swimmer might be reached if caught in a hydraulic below.

Helicopters may also be used for vertical rescue, but these too must be considered a last resort, for reasons that will be discussed later in this chapter.

Bridge Lowers
A good example of a successful bridge lower is a rescue at Powerhouse Rapid on the Ocoee River in the fall of 1982. A

paddler broached his open canoe on the central bridge support. As the boat wrapped, his left foot became entangled in the webbing of the thighstraps. He could not free himself, although he was able to breathe periodically by doing pull-ups on the gunnel. The force of the water made this difficult and he was tiring rapidly.

By good fortune, a commercial raft trip was just upstream and quickly responded to the situation. A rescuer tied herself onto a rope and climbed down the iron bridge girders to the victim. The rope was simply run over the bridge railing and the rescuer belayed by fellow guides. The rescuer's first action was to tie off the victim with another line from above to stabilize his position and keep his head out of the water. Next she cut the tangled thighstrap to release the paddler, who by this time was becoming disoriented from fatigue and incipient hypothermia, and moved him onto a small shelf on the lower part of the bridge piling. She then rigged a leg-loop harness for the paddler, and he was pulled to the top of the bridge by sheer manpower. The elapsed time of the rescue was about fifteen minutes.

The rescuers could have rigged a Telfer lower in this case, but a direct vertical lower was faster and more expedient, since there were plenty of people available to haul on the rope. As in most river rescues, speed was the overriding consideration. Fortunately, the rescuers were professional guides used to working together, and this made organization of the rescue much easier.

When organizing a bridge lower, consider the following:

- The rescuer to be lowered should be one of the lightest and most athletic of the party. He should have at least a knife, a helmet, and a lifejacket.
- A simple harness can be fashioned from a rope or a piece of webbing (Fig. 7.5). It is not particularly comfortable, but it is quick to tie and is *relatively* secure for the rescuer.
- A chest harness can be used in addition to a sit harness for increased safety. Using the two in combination is much more comfortable than either alone.
- Backup safety ropes and boats should be positioned

Fig. 7.1. A direct bridge lower on the Ocoee. The victim is given a rope loop and pulled up by simple manpower. A backup belay line would have been a good idea. (Photo by Helen Mary Johnson)

Fig. 7.2. Three steps in tying a sit harness. First, make a loop of rope as long as the distance from one shoulder to the opposite outstretched finger, and tie it with a double fisherman's knot.

Fig. 7.3. Then pull it into three loops, one around each hip and one under the crotch.

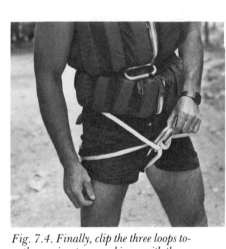

Fig. 7.4. Finally, clip the three loops together, using two carabiners with the gates reversed.

downstream in case the victim or the rescuer falls in.

- Part of the bridge structure itself, such as a railing, can often be used with a friction belay to make the lowering smoother and slower. Wrap the rope around the structure a few times, according to the size and shape of that part of the bridge. One danger of this is that the rope can be weakened or cut by a sharp edge. A safer lowering system is the carabiner brake system (Fig. 7.7). The more carabiners there are in the system, the greater the friction. Note that the carabiners are doubled and the gates are on opposite sides and reversed. This takes more time and equipment to set up than a simple friction lower.

- Using double lines for the lowering reduces the chance of rope failure. It also increases friction, which means a slower, more controlled descent. For added safety, add an independent line to serve as a belay.

- If possible, set up the carabiner brake system or friction wrap on the downstream side of the bridge. Since accidents almost always happen on the upstream side of a bridge, this gives the rescue team the width of the bridge to work in. If the rope must pass on the up-

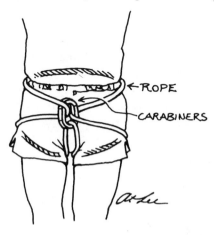

Fig. 7.5. A simple sit harness.

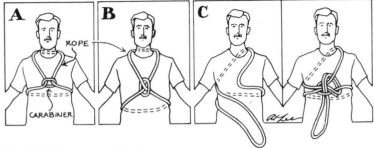

Fig. 7.6. Three types of chest harness.

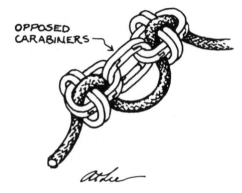

Fig. 7.7. The carabiner brake system.

stream side of the bridge, make sure it is padded or run through a slung carabiner to reduce friction.

Once freed, both victim and rescuer must be raised to safety. Unless the patient needs constant attention, he should be raised first. If enough people are available, have them pull the rope up over the railing, hand over hand, but set up an independent belay in case the main

Figs. 7.8-7.10. The bridge lower. This rescuer uses the sit harness shown above. The lowering rope is run through a slung pulley (actually two carabiners with the gates reversed) on the bridge structure. She is lowered by running the rope around the bridge railing to provide enough friction for a smooth descent. Ideally there should be a backup belay rope. She carries an auxiliary line with her.

Fig. 7.11. A carabiner brake. If no convenient belay point is available this system can be used to provide enough friction for a safe lower.

Fig. 7.8

Fig. 7.9

Fig. 7.10

Fig. 7.11

haul line fails. A carabiner brake system will work too, but it is very slow to work the rope back through it (i.e., back upwards). It is much easier to transfer the load to prusiks, remove the carabiners, and then use a friction belay on the bridge.

If there are not enough people for a hand-over-hand haul, consider a Z-drag system. An automobile or winch might also provide the necessary power.

Tyroleans

We use the word "Tyrolean" to describe any kind of overhead rescue that uses a fixed anchor line stretched across the river. It is a method adapted from the standard Tyrolean traverse technique used in mountaineering to cross crevasses or to travel from one rock spire to another.

Some see the Tyrolean as a system which will magically snatch the victim from the water. This may indeed be the case under some conditions, but for most purposes the Tyrolean is only the first step in a shore-based rescue: its purpose is to get the rescuer near the victim. In most instances, the rescuer will use his position to attach auxiliary lines to the victim or the boat so that the extrication or recovery can take place from shore.

The Tyrolean rescue is dangerous to the rescuer: if the rope or another component breaks, or if an anchor point fails, it means a fall onto the rocks or into the water. Even if the water cushions the fall the rescuer may still become entangled in the ropes. Because of the danger to the rescuer, the time required to set it up, and the amount of equipment and manpower involved, the Tyrolean is another technique of last resort and should be considered only when other methods have failed or are not feasible.

Professional rescuers have the luxury of sewn sit harnesses, steel cables, winches, and various other tools of the trade. Recreational paddlers will have only those pieces of rescue gear they normally carry with them, which again emphasizes the importance of every paddler carrying a rope and some basic rescue gear. Tyroleans, more than any of the techniques mentioned so far, use a lot of rope.

Some general considerations for using Tyroleans are:

- The safety of the rescuer should be a primary concern. Consider the consequences of failure: if you have enough rope, double your anchor lines and use a separate belay for the rescuer, if possible, so that he can be brought to shore if the anchor line fails.
- The quality of the rope you use for the anchor line is important, since the safety of the rescuer depends on it. Climbing rope is ideal but will seldom be available. The ½-inch polypropylene rope described in Chapter 2 is better than the ⅜-inch rope in most throw bags (which should be doubled if possible). The rope you choose for the anchor line should be of the highest possible quality and should not have any splices in it.
- Operating a Tyrolean, especially climbing up and down, is quite strenuous. The person selected to do this should be light and athletic. He should be equipped with at least a helmet, a lifejacket, and a knife.
- Place the anchor line under tension by hand if possible: knots and Z-drags reduce the strength of the rope. Use a Z-drag only if you cannot place the anchor points

Fig. 7.12. Climbing a tree using girth hitches.

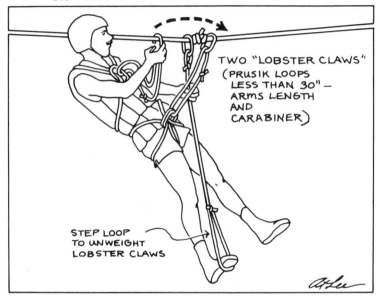

TWO "LOBSTER CLAWS"
(PRUSIK LOOPS
LESS THAN 30" –
ARMS LENGTH
AND
CARABINER)

STEP LOOP
TO UNWEIGHT
LOBSTER CLAWS

Fig. 7.13. How to pass a knot on an anchor line. This rescuer uses one lobster claw clipped into a chest harness and another into a sit harness, making for a much more comfortable arrangement than either alone.

high enough to prevent the rope from sagging into the water under the weight of the rescuer. If possible, back up the anchor line under tension with a slack anchor line.

- Trees usually make the best anchor points for the anchor line. If you can pass the anchor line over a large branch or through a slung carabiner, you will be able to bring the rope down to ground level, where it can be tensioned and secured by hand. Rather than tying off the rope, it is often a good idea to assign a team member to hold the rope in his hands — having wrapped it several times around a tree. This makes it easy to release if the rescuer gets into trouble. A knife will do the same thing, of course, but it is harder on the rope.

- Position backup safety ropes and boats downstream in case the rescuer goes in or the victim is freed suddenly.

- You can climb trees without low branches by using a

system of three slings girth-hitched around the trunk: one sling for each foot and one for a chest sling (Fig. 7.12).

It may be necessary to join two or more ropes to make the anchor line (use a double fisherman's knot), and if so the rescuer may have to get his harness carabiner past the knot. The best way to do this is to use a small sling the rescuer can step into to get his weight off the harness carabiner, unclip himself, and clip himself in again beyond the knot. Use a second carabiner and tether to keep the rescuer from falling.

There are various types of Tyrolean rescue:

Ranger Crawl   This is the simplest form of the Tyrolean, and the most dangerous to the rescuer. It is quick to set up, and it requires a minimum of equipment: a single anchor line upon which the rescuer crawls. Another version, which provides slightly more security, is a double anchor line made with parallel ropes (Figs. 7.14-7.15). In both cases the anchor line(s) is (are) slowly released to lower the rescuer into position, which means the rescuer is faced with an uphill crawl to get back to shore. Use the ranger crawl only when falling in means no more than getting wet.

Sliding Seat   A simple rope seat attached to a single anchor line by two carabiners provides a fairly comfortable way to slide along (Fig. 7.16). The rescuer slides himself along the anchor line until he reaches the accident site, and the line is then lowered to let him make contact with the victim.

Self-Lowering-Rescuer System   After making himself a harness, the rescuer rigs a Z-drag from the harness to the anchor line (Fig. 7.17). The rope is tied to the harness, passes through one carabiner on the anchor line, down through a carabiner on the harness, up through a second carabiner on the anchor line, and then back down to the rescuer's hands. During the rescue the rescuer adjusts the Z-drag so that he is within arm's reach of the anchor line, holding the loose end of the rope in check with

Figs. 7.14-7.15. Two versions of the ranger crawl. One uses a single rope, the other a double.

several (four to six) wraps around one thigh. He then pulls himself out over the accident site, hand over hand, with his body weight supported by the harness. When he reaches the accident site he removes the thigh wraps, lowers himself in an action very similar to rapelling, then rewraps the leg to hold himself in place. When he has finished, he can raise himself again, using the Z-drag, and slide back to the shore. If he falls into the water the Z-drag should run free. This is a good system if there are only a few rescuers.

Shore-Based Lowering System   This is similar to the previous method, except that the lowering is done from shore using manpower, a friction wrap, or a carabiner brake system. Attach a carabiner or pulley to the anchor line, holding it in place with a tag line from the opposite shore or with a prusik set on the anchor line (Fig. 7.18). This system allows the rescuer to concentrate on the victim, although it does require more people to raise the rescuer and return him to shore.

Fig. 7.16. The sliding seat is a quick and convenient way to get out over the water, although not nearly as flexible as some of the other systems. This confused rescuer has his lifejacket on upside down.

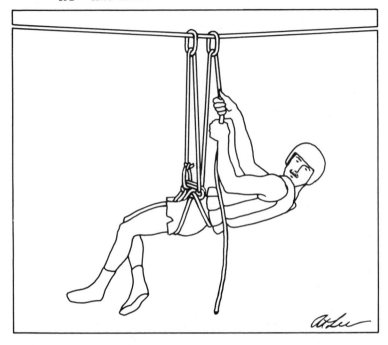

Fig. 7.17. The self-lowering-rescuer system.

Rope-Ladder Rescue Probably the most versatile of the Tyroleans, this version includes a rope ladder for the rescuer to climb up and down on. The ladder itself is made by tying a series of double figure-of-eight knots. Each loop should be big enough to step into and should overlap the next one (Fig. 7.19). The rescuer clips directly onto the anchor line, pulls himself out over the accident site, clips the ladder onto the anchor line, and then climbs down the ladder to the victim. A prusik tied to the anchor line keeps the ladder in position. The rescuer's harness has two short tethers (usually called lobster claws) with carabiners on each. As he climbs down, he must always keep one of these clipped into a loop of the ladder. This system might also allow a victim to climb on the ladder and be pulled to shore.

Helicopter Rescue

Helicopters are wonderful tools for rescue and evacuation — if used correctly. Like anything else, however, they are subject to physical laws and have their limitations, and many of the problems that arise in helicopter rescues stem from ignorance of those limitations. Most people, and this includes many rescue professionals, are not familiar with the capabilities of helicopters.

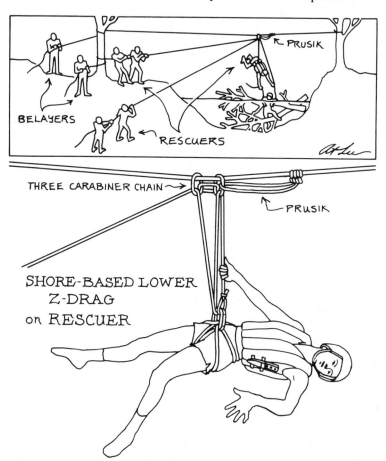

Fig. 7.18. The shore-based lowering system.

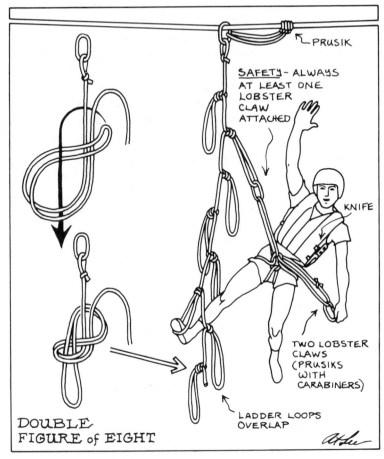

PRUSIK

SAFETY – ALWAYS
AT LEAST ONE
LOBSTER
CLAW
ATTACHED

KNIFE

TWO LOBSTER
CLAWS
(PRUSIKS
WITH
CARABINERS)

LADDER LOOPS
OVERLAP

DOUBLE
FIGURE of EIGHT

Fig. 7.19. The rope-ladder rescue.

The advantages of the helicopter are mobility and speed. It is sometimes faster than other systems and is especially good for cutting evacuation time to a minimum. It can hover over an accident site where there are no nearby bridge pilings or anchor points and can rescue victims from the middle of rivers too wide for ropes to reach across. Some helicopters, especially military ones, are also equipped with powerful winches, which can be used for vertical rescue.

Fig. 7.20. When making the rope ladder, be sure to tie the loops close enough together. This is a good way to measure the distance.

The disadvantages of the helicopter are many and should be carefully considered before use. They are fast in flight but are often slow to respond. Even in areas where they are used frequently, the tasks of getting the required clearances, scrambling the crew, and seeing to other administrative details may use time precious to a river rescue. Once the helicopter is on the scene there are often communications problems: unless the group on the river has a radio (unlikely) and can use it to talk to the helicopter (even more unlikely), they must use ground-to-air signals, which are slow and inexact (see Appendix G). Helicopters also need a lot of room to operate, which limits their usefulness in narrow, tree-lined river gorges. If a helicopter snags a rotor and crashes, an already tragic situation becomes much worse. Furthermore, the rotor backwash from a helicopter hovering over a victim can dramatically increase the possibility of hypothermia.

There is an almost universal tendency to overestimate the capabilities of helicopters, which has led rescuers to try to lift water-filled kayaks or unpin rafts with them. Probably the worst problem with the use of helicopters in river rescue, though, is the

Fig. 7.21. A helicopter can hover over an accident site where there are no nearby anchor points and can rescue victims from rivers too wide for ropes to reach across. (Photo courtesy of Bell Helicopter)

temptation to put off rescue attempts "because the chopper will be here any minute." Rescue attempts on the river *must not be put off for any reason*, unless it is absolutely clear that there is no other alternative.

In May 1980 a kayaker became pinned on the upstream side of a boulder on the Kern River in California. He could breathe, but he could not get out of the boat. There were several delays in assessing the situation, and it was nearly half an hour before help was summoned. A deputy sheriff on the scene concluded that a shore-based rescue was too dangerous and prohibited a group of commercial rafters from making any attempts. One guide and a private kayaker did swim out anyway and hold the victim's head up.

It was crucial to the deputy's thinking that he had radioed for a helicopter and expected it to be there within 15 minutes. He did not know that, as Charlie Walbridge comments in *The Best of the River Safety Task Force Newsletter,* "it would take 25 minutes to get the necessary clearances and 35 minutes to scramble the crew before the 15-minute flight could commence." Nor could he communicate with the helicopter's radio, which was on a military frequency.

What followed was a near tragedy. By the time the helicopter arrived the victim had been in the water nearly two and a half hours and was becoming hypothermic. The helicopter had been called one and a half hours before. The helicopter lowered a rescue collar to the victim, who was still in his kayak, and tried to lift both to the shore. It lost lift and almost crashed in the process. Fortunately, in spite of this error and more bungling in treatment and evacuation, the injured kayaker escaped from both accident and rescue with only minor bruises.

Walbridge comments that "the people in the helicopter had no idea how much water was in the kayak, or if they had enough lift to do the job, much less if this could be done without injuring the victim." He concludes that "river rescues are not like other kinds. Time works against you and a helicopter is, at best, a backup."

A similar incident in New Zealand was described to us by our friend Bob Karls, who was working for a commercial rafting outfit there at the time. One of the outfitter's rafts was pinned on a rock on the Shotover River. Despite Bob's offers to set up a Z-drag, the outfitter decided to use a helicopter to unpin the raft. "They use helicopters for everything down there," says Bob, "and they didn't see why they couldn't use it for this."

The outfitter tied the ropes from the helicopter to the raft's D-rings (in spite of Bob's advice to tie them to the thwarts) and the chopper began pulling. "Just as it looked as if the raft might move," Bob recalls, "several of the D-rings popped loose. The ropes to them were stretched like rubber bands and they jerked the metal D-rings and ropes right into the rotor blades. And I was right under it." By a miracle the helicopter managed to land without crashing.

The lesson of these incidents is clear: communication, capabilities, and time must all be considered. Does this mean that helicopters should never be used? No. Helicopters are routinely used in the Grand Canyon for evacuation, and the following incident on the Cheat River in West Virginia (reported in *The Best of the River Safety Task Force Newsletter*) shows how they may be used to advantage.

An outfitter's raft flipped on the Cheat at high water. Everyone was rescued except one person, who ended up stranded on a mid-stream rock. The river was already beyond the "safe" level for trips and rising. The victim was shaken, and although the outfitter was able to get a kayak over to the rock a huge strainer downstream made the ferry a real risk for a raft. The river was too wide and swift for a Telfer or Tyrolean system, but it was relatively unobstructed. The local rescue squad called a helicopter, which was able to execute a single-skid landing on the rock and pick up the victim. In this case the helicopter, which was already in the area, was the best alternative.

Helicopter Evacuation

On the whole, helicopters are better used for evacuation, when their speed may be used to get a patient to a hospital quickly, than for rescue. However, even if a helicopter has been called the rescue and evacuation should proceed as if it were not available. If the helicopter is not needed in the rescue, the rescuers should look for a suitable landing zone for an evacuation. It is much safer, unless the patient's injury is so serious that he can't be moved, to use a proper landing area than to force the pilot into making a dangerous pick-up at the accident site.

The Landing Zone   Set up the landing zone in a flat, clear space with no obstructions around it. Telephone wires and power lines are a special hazard, since they are hard for the pilot to see. An isolated knob or hill is ideal. The landing area itself should be flat, however, since helicopters cannot land on a slope much greater than 8 to 10 degrees. The landing zone for a standard UH–1 "Huey" series helicopter should be a minimum of 35 paces wide. A pilot can set a helicopter down in a circular area this size, but he will find it difficult to take off or turn around. The landing zone should therefore be three or four times as long as it is wide, and the long axis should face into the wind (Fig. 7.22). If there are

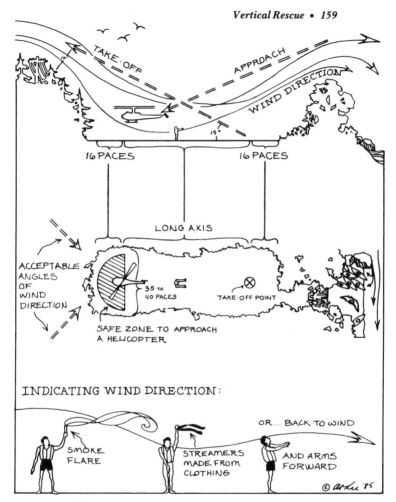

Fig. 7.22. Helicopter landing zone.

obstacles like trees and power lines at the end of the "runway," this distance must be greater. While helicopters *can* take off and land vertically, it is much easier for them to take off into the wind, just as it is for airplanes.

Clear the landing zone of all loose debris and brush, leaving no obstacles higher than one foot. Mark it clearly and if possible arrange for some indication of the direction of the wind. Paddle jackets or brightly colored clothing laid in a T or H pattern will serve as markers, but you will have to stake or weigh them down to keep them from blowing away. Indicate wind direction with smoke or a streamer made from clothing or by standing with your back to the wind and holding your arms forward.

Evacuation Techniques   Do not rush out to the helicopter when it lands: the pilot may want to move it around first. Instead, make sure you have the attention of the pilot and wait until he indicates he's ready. Approach the helicopter *only* from the front or side — stay away from the tail rotor. If the ground slopes, approach from the downhill side only and duck as you pass under the arc of the rotors. Once you are at the helicopter, a crew member will instruct you where to put the patient. If it is a medivac ship there will be a doctor or EMT aboard; if not, a first aider should accompany the patient to the hospital, if there is room on the aircraft.

Conclusion

Tim Setnicka, who has worked extensively with rescue helicopters in Yosemite National Park, reminds us in *Wilderness Search and Rescue* (Appalachian Mountain Club, Boston, 1980) that rescuers "must always consider the negative impact of any one of these factors: bad weather, malfunction, darkness. We therefore want to stay *helicopter independent* in [search-and-rescue] planning and thinking, in spite of the seductiveness of constant reliance on [helicopter] support."

Figs. 7.23-7.28. Ocoee Bridge rescue, October 1982 (mentioned at the beginning of this chapter). John Norton, an Atlanta canoeist, was rescued by Karen Berry and a group of river guides from High Country, an Atlanta-based outfitter. (Photos by Jeff Ward)

Fig. 7.23. Norton's left foot became entangled in the thighstrap webbing of his canoe as it broached on the bridge piling at Powerhouse Rapid. He could breathe only by pulling up on the gunnels.

Fig. 7.24. The rescuers first attempted to free Norton by lowering a rope to him. When this failed they lowered Berry to the broached canoe.

Fig. 7.25. First Berry tied a stabilization line to Norton's wrist, to ensure that his head would remain above water.

Fig. 7.26. Then she cut the thighstrap, releasing Norton . . .

Fig. 7.27. . . .who was then pulled up on a shelf of the bridge piling in the eddy behind his pinned canoe.

Fig. 7.28. Finally, Berry rigged a sit harness for Norton, who was then raised by simple manpower.

There is no right or wrong way to perform a rescue. What matters is whether it is successful.

LES BECHDEL

· 8 ·
Organization for Rescue

The first few moments of rescue — from the time the need for rescue is recognized until the rescue actually begins — are critical. The most important part of that time is devoted to the organization of the rescue. Because of the very limited time available — a minute or less in a "head-down" pin — organization and timing are vital.

The rescue leader must take charge of the situation *immediately*, assigning responsibilities and directing rescue efforts. If the crisis should involve the rescue leader, the leadership role must be assumed by the most experienced paddler in the group. Reaction time is critical: seconds and tenths of seconds count. This is no place for arguments or democratic discussions: the directions of the leader must be followed. For these reasons we suggest that an organized group take the following steps. We realize that some will condemn us as safety fanatics because we are trying to organize a very individualistic sport, but because of the shortness of time available in most rescues there is no substitute for organization and leadership.

The Rescue Process

What factors should you consider when organizing a rescue? You must assess the situation, communicate the problem, determine the best method of rescue, and organize the rescue.

Assess the Situation

- How many people are involved? Where are they?
- Where is the victim? Is he entrapped?
- Can the victim breathe? Is it a "head-up" or a "head-down" entrapment?
- Can the situation get worse? How? Can the boat shift? Is the water rising? Is hypothermia a consideration?
- How much time do you have? Consider the time left for the victim, the hours of daylight remaining, and the time needed to summon help.

Communicate the Problem

- To the victim. What is the rescue plan? The person being rescued will need reassurance and will certainly want to know what is being done to help.
- To others in your party and to other parties on the river as quickly as possible, so that everyone knows what the problem is.
- To outside resources, such as a dam keeper or emergency medical services.

Determine the Best Method of Rescue

- Consider the time available to set up and use the rescue system. This is vital if the victim cannot breathe or is in extreme danger.
- What danger will each method pose to the rescuers? This question is too often overlooked. A reasonable risk might be accepted to save a life, but beware of adding another victim.
- How many people will be needed? With some methods, untrained bystanders can assist.
- How much equipment is required? Some otherwise desirable method may have to be eliminated if critical hardware is not available.
- What is the probability of success? Is the method appropriate to the problem?
- Can different methods be used concurrently or as a backup? A quick-to-use method like a boat-based res-

cue can be tried first while another team readies a slower method like a Tyrolean.

Organize the Rescue

- Appoint individuals to rescue teams if enough people are available. Designate team leaders and tasks. In small groups, each paddler will form a team of one.
- The most important thing is to stabilize the patient's condition and start the extrication. The most skilled paddlers and river-wise people should be appointed to the extrication team.
- While the extrication is in progress, prepare for first aid, CPR, and hypothermia treatment as necessary.
- Send for help. Notify local search-and-rescue teams. Prepare for ambulance pick-up at the nearest road. If you are on a dam-controlled river, notify the authorities to shut off the water. Call the nearest hospital and give a description of the injuries, so that proper medical facilities can be made ready.

That's a lot to think about, isn't it? How much time do you have to get going? Because of the life-threatening nature of many river emergencies, you must be able to begin the rescue *within 15 seconds* of the time of the accident. Experience, river sense, and presence of mind count for a lot here: if you have been following the course of events leading up to the accident and automatically analyzing them (the "what if?" factor) you should be propelled into action almost instinctively. But, while we cannot overemphasize the importance of speed in a river rescue, there is no substitute for a moment of reflection before starting: it is no rescue at all to hurry into an ill-considered attempt that endangers the lives of the rescuers and is doomed to failure because of poor and hasty organization.

Leadership

As a rescue leader you must concern yourself with clear, unemotional thinking. This is not always easy, particularly if the safety of a friend is involved. If possible, you should not become

Fig. 8.1. Rescue-team organization is vital for success. Here the rescue leader controls the efforts of two rope teams during a snag-tag exercise.

directly involved with any of the individual tasks of the rescue. It is important to have someone who can oversee the whole operation and not get too involved in any one problem. Obviously there will be times when you can't do this: you may be the only one qualified in a certain skill, for example. Remember that your duties as a rescue leader are to *observe, organize, and direct* the efforts of the entire rescue team.

Methods of Rescue

One of the most important factors on the river is always time — the time it takes to set up and use a rescue system and the time a victim may have left. Almost inevitably, time must be traded off against the danger to which a rescuer must be exposed. If a victim's remaining time is measured in minutes or even seconds, a rescuer may be prepared to accept some danger to himself in order to save his life. But this trade-off must be considered carefully and the attempt skillfully made, with all possible safety precautions taken for the rescuer, or he could end up as another victim.

Another factor is the use of the appropriate system, which means rescuers must be familiar with many different rescue systems. Rescuers must also consider the available manpower and equipment. Many systems can be used concurrently or as backups to each other.

Generally, if the situation is immediately life-threatening, the fastest system consistent with the rescuer's safety must be used, with safer but slower systems as backups; if people must be rescued but are not in immediate danger, slower and safer systems should be used; if only equipment is involved, choose the safest system possible for the rescuers.

As mentioned in Chapter 5, shore-based systems are usually safer for the rescuers than boat-based systems. Some systems, like the strong-swimmer rescues, are quick to set up but dangerous to the rescuer. Others, like the Tyroleans, are both slow *and* dangerous. The choice is an important one, and will often weigh heavily on the minds of the rescuers.

Team Organization

It would be impossible to set out in these pages exactly what organization you will use for a rescue. Each rescue is different, and the equipment, experience, and numbers of people available will vary greatly. One primary goal is to make the best use of all available manpower, and the most effective way to do this is to organize people into teams. Our intention here is to provide a guide to all the teams you might have in the best of all possible worlds. In a small group some or all of these team functions will be performed by individuals; in some cases one person may have to do them all.

The organizing should be done by the rescue leader, except when he is deeply involved in an extrication or other rescue task (for which he might be the only one qualified) or in the case of a small group of experienced paddlers who, thanks to experience and instinct, react instantly. If for any reason the rescue leader cannot organize the rescue, someone else, preferably the next most experienced person, must take over.

The following teams are usually needed:

- *Extrication team* Most needed in a pin or entrapment, more often than not this team is not appointed at all but consists of the people nearest to the scene of the accident. Because speed and efficiency are critical, this team should comprise the most experienced and water-wise paddlers.

- *First aid team* The "aid" team prepares to treat the patient while the extrication is going on. It should include medically the best-trained people present and is responsible for the treatment of the patient. The aid team must determine where CPR will be performed, who will do the compressions and who the breathing, what first aid is needed and where, and whether a litter or backboard will be necessary.

- *Evacuation team* The "evac" team coordinates with the aid team to determine the best method of evacuation, then selects the evacuation route. The team also con-structs or locates litters or backboards and performs the evacuation.

- *Communication team:* The "commo" team goes for help. After receiving instructions from the trip leader, the team will go by the best means available (foot, paddling, hitch-hiking) to make contact with outside agencies such as search-and-rescue teams, ambulances, and hospitals. Contingency plans should be discussed *before* leaving the other teams. It is a good idea if possible to send people out in pairs, in case one gets injured.

- *Support team* The support team takes care of every-thing else. Someone must look after the other group members, who may be having their own problems with shock, grief, or hypothermia. If the group needs to keep moving down the river or split up, who will be the

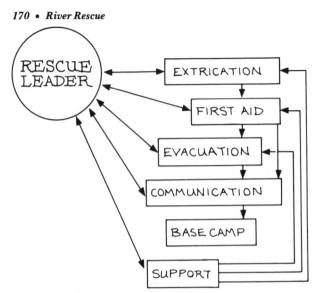

Fig. 8.2. Rescue organization.

new leader? The support team may have to prepare a fire, a meal, or a campsite. If the accident happened near a road someone will need to control curious on-lookers. One member of the support team should be stationed upstream of the accident site to warn other paddlers; another should stand downstream with a backup rope for the extrication team. Finally, some-one should be taking notes and, if possible, photo-graphs so that a record of the incident can be made.

- *Base camp* It may be necessary to set up a base camp with someone to receive and coordinate information and resources, comfort friends and relatives as needed, provide continued support with food and clothing, and deal with the media. Often this function is per-formed by the commo team at the take-out or put-in.

If the First Attempt Fails

What happens when the first attempt fails? Obviously you cannot just give up while you have the means in your power to rescue someone.

If enough people are available, some should be working on an alternative method of rescue while the primary method is being tried. For instance, if a paddler is pinned in a boat in mid-stream, you might first try sending a rescue team out in a canoe or raft. While this is being done a second team can be stretching a rope across the river to prepare for a Telfer lower or a Tyrolean rescue.

Reassess the situation, using the same criteria as the first time. Don't give up! Was there something you missed? Has the situation changed? Are there other methods you haven't considered? Solicit suggestions from others. Are the rescuers getting tired or otherwise endangering themselves? In your concern for the victim do not ignore the effects of hypothermia or shock on the rescue party. Has someone been sent to notify others, such as a search-and-rescue team, who might have equipment you don't have?

Keep trying, and keep thinking. Keep the rescuers safe, but don't give up until the situation is clearly hopeless. Remember that apparent victims of drowning have been revived after over half an hour's immersion in cold water.

Thinking the Unthinkable — The Failed Rescue

In a book, all rescue attempts can be successful, but in real life you must be prepared to deal with those that fail. An unsuccessful rescue means someone is dead. Death is not a pleasant thing to think about, but it's something that you must consider as a logical extension of organizing for rescue.

As we have said, the rescuers must keep trying to save the victim until the situation is clearly hopeless. What then? If there is no doubt but that the victim is beyond help, a decision must be made as to whether to continue efforts to recover the body. Because of the emotional attachment of people in the rescue party to the victim, this may be a hard decision to make, but you should continue recovery efforts only if they are not compromising the safety of the rescuers.

Consider the condition of the rescuers. They are almost certainly tired and emotionally upset by what has happened. Is the party near a road or take-out? How much daylight is left? Should the party continue or abandon the trip? The safety of the living must come first. If the group is near a road or civilized area and recovery attempts have been fruitless, it may be better to leave the attempt to a sheriff's department or search-and-rescue unit. If you are on a multi-day wilderness trip, the victim may have to be left for a later search party. It is well-known among mountaineers that a party will tend to abandon an expedition after a death, regardless of the actual difficulties. Certainly the death of a companion casts a pall over the whole undertaking, but the risks and rewards of continuing must be evaluated realistically.

Suppose the rescue is unsuccessful and the victim has been recovered. What do you do with a dead person, especially a friend? If all attempts at revival have failed, remove the body to a safe place until you can either evacuate it yourselves or arrange for it to be done. If you are in the wilderness and far from help you may have to bury it, either temporarily or permanently. In either case the grave must be marked.

In most places in the United States a person must be pronounced dead by a civil official, usually a coroner or a medical examiner. The coroner must file a report and, if he thinks the circumstances warrant it, order an investigation. Any death must be reported to the local police department or county sheriff, along with the circumstances. They will want specific information. To be able to provide this information, the people involved in a rescue, whether it involves a fatality or even a near miss, should debrief as a group as soon as possible. It is surprising how quickly people forget details and how subjective their impressions of time are. The only way to be sure is to write things down as soon as possible after the event. Try to specify exactly what happened: who did what when and in what order. If possible, take photographs of the accident site, the rescue, the evacuation, and the victim. It may seem morbid to take pictures at a time like this, but they are invaluable when trying to piece together events later. A photograph will not forget details.

Another reason for setting all this down is to pass on the experience, good or bad, to others who might at some time be in a similar situation. Write down a summary of the accident and send it into the River Safety Task Force (see Afterword).

Liability

In these days of the "sue everybody" society the question of liability frequently comes up. Unfortunately, there is a lot of misinformation about it, even in the legal community. In the simplest terms, you are liable to a person when, through fault or omission, you cause injury to that person or his property. After that it gets complicated, and discussions of this subject have filled many law books.

What happens when you see someone in trouble on the river? Are you under an obligation to help? and what happens when you do? Many states have "Good Samaritan" laws, which protect a would-be rescuer who acts in good faith. Some even make it a crime *not* to render aid. What is good faith? It just means trying to do the right thing within the limits of your knowledge and capabilities.

Still, it is impossible to have a lawyer on every trip, and in some cases the river you are on may even mark the border between two states with different liability laws. As a practical matter, there are already too many things to think about during a rescue: you don't need to add liability to the list. Your knowledge will protect you: if you know river rescue and first aid, use good judgment, and try to prevent accidents before they happen, your chances of being sued as a private boater are very slight.

Reactions

We have talked about successful and unsuccessful rescues and some of the administrative considerations involved in both, but what really happens in your mind when there is a fatal accident on the river?

At the beginning of the emergency there is faith, a trust in the system, a certainty that the victim will be saved. This is replaced by disbelief that the accident is really happening, then by hope that a miracle will occur. Inevitably, there is the realization that the accident actually has happened and the anger and frustration that go with that realization. This is often coupled with a strong sense of personal guilt, regardless of the circumstances. Finally, there is the acceptance of death, which leaves a numbing feeling that lasts for some time.

Fig. 8.3. Rescues save lives, but sometimes they fail. What would your reactions be to the death or serious injury of one of your party? (Photo by Cindin Carroll/ Nantahala Outdoor Center)

As a conclusion to this chapter, here is Les describing some of his feelings after a drowning death on the Bio-Bio River in Chile.

"It was to be my last trip on the Bio-Bio. After four seasons there I had come to love the beauty of this rugged river and its whitewater. We were running Lost Yak rapid and my raft was first. We took on a lot of water and went through the next rapid, Lava South, as well. As I returned to Lost Yak on foot, another raft was coming through and I saw Ted in the water away from the raft. Because of the width of the river we were unable to use safety ropes to reach him, but John approached him in a kayak. Ted grabbed the boat and they began to struggle toward shore. I had *faith* in John's ability, but a lot of cold, fast Class IV water lay between them and safety.

"My mind flashed back to a similar incident two years before [the rescue described in Chapter 1]: we were at the bottom of Lava South and a person from another company, who had fallen out of his raft at Lost Yak, was forced to swim through Lava South. That time we were lucky to be in the right place at the right time.

"Ted died only a few feet from shore. Caught under a rock, he was still clutching the kayak's grab loop. *Disbelief.* This can't be happening. I watched helplessly from the opposite shore as Ted disappeared face down into Lava South. John had lost his paddle in the struggle to free Ted and was trying to follow him on foot. All our other rescue attempts, some foolish and some brave, ended without success.

"Somewhere during all this my hopes crashed. Frustration had been growing during the pursuit of Ted's body and it ended in *anger.* It wasn't fair. We had done everything by the book, taken every precaution. He was an experienced paddler. I had to turn my anger and despair into something else. We had to start thinking about ourselves. As I walked back to the rest of the group, waves of *guilt* washed over me. We should have portaged around Lost Yak. I should have been able to do something to save Ted. It was my fault he was here. And now he was dead.

"Getting the group together that night wasn't easy. Each of us had our own way of dealing with the grief. We didn't know whether to quit or go on. We knew we had to get out of the canyon; after that we would take it one day at a time. We finished the trip and took out four days later at Santa Barbara. The authorities found the body the next day.

"My trip lasted longer. We sent Ted's body back to West Virginia, and it was there at his funeral that I *accepted* the reality of his death. His mother asked me if the river was beautiful and I said that it was. She said, 'At least he died doing what he loved the most . . . paddling whitewater on a beautiful river.' "

Keep strong, if possible. In any case, keep cool.
B.H. LIDDELL HART

· 9 ·
Patient Care and Evacuation Techniques

"We had put together a good rescue and I was congratulating myself as we neared the end of the evacuation trail. Then, as we slid the litter over a fallen tree, the patient cut his finger and started to go shocky on us.

"He had broken his leg in Corkscrew Rapid on the Chattooga. The rescue was quick and minutes later we had him on shore with an air splint in place. He was obviously in pain but was staying in control. We got the emergency backboard, and within half an hour we were carrying him down the rocky banks of the river. What we failed to do was instruct our patient on how to protect himself while being transported in a litter. When we slid him over the log we hadn't noticed that he was gripping the edge of the litter. The injury to his finger sent him over the threshold of tolerance and he began losing the stoic control he had been exercising. His respiration became shallow, he wouldn't respond to our comments, and he had poor color. He was at the point of going into shock. With great effort he brought himself back, but it was close, too close. We were lucky and it was another lesson in patient care."

Until this point our concern has been with rescuing the victim of an accident. At the moment of rescue, however, a victim becomes a patient, and this chapter is concerned with dealing with the person who is now your patient. It is not our intent to cover first aid procedures, but we do want to highlight some thoughts on post-rescue care that should concern any rescuer.

Anybody involved with outdoor sports should have some form of first aid training. Outdoorspeople frequently find themselves engaged in an active sport where injury in a wilderness setting is a distinct possibility. Professional help can be hours, days, or even weeks away, which means some training in dealing with medical emergencies is a must.

The Initial Contact

A contact rescue is a hands-on approach by the rescuer. There is no rope between victim and rescuer. This is usually the only way to rescue an unconscious victim. When making a contact rescue, you must be wary of the conscious victim: he may be in severe pain or have just had a close brush with death. A distraught victim in the water may make a desperate lunge for the rescuer or try to climb on top of him. A victim stranded on boulders in midstream, pinned in a boat, or with his foot entrapped may have been in that position for some time: his actions may be difficult to anticipate.

When approaching a victim it is a good idea to have a rescue plan in mind. Communicate it to him. Tell him exactly what you are doing and planning to do, and what you want him to do. Ask his name. Ask about injuries: "Where does it hurt?" If you are new to the scene, ask if there are others in the party. Try to get as much information about the accident and the patient's condition as you can. If the patient is conscious but not responding to your inquiries, be cautious.

If you have witnessed the accident and know the patient, use his name. Be calm and reassuring. Be honest about the situation, remain positive, and try to inspire trust and confidence. The first few seconds of interaction may make the difference between a patient who remains cool and one who goes off the deep end. As we saw in the opening paragraphs of this chapter, a patient's mental condition is critical and should be monitored at all times. An otherwise insignificant injury or even a seemingly innocuous remark ("God, he looks terrible!") overheard by the patient may be enough to send him into shock.

If the patient is stationary and out of immediate danger, your primary concern should be to stabilize his condition so that it doesn't get worse. We have already discussed strategies to use in pinnings and entrapments, so the emphasis here is on assessing the nature of the patient's injuries and his emotional state. Do not aggravate the patient's condition by hasty, unnecessary movement. Make certain that the patient is breathing and not bleeding.

A contact rescue that occurs in the water does not offer much time for a survey of the patient's condition. If the victim is conscious you can ask about his injuries; if the victim is unconscious and breathing, stay on the upstream side of him and keep his head above water. In heavy water, position yourself under the patient with his and your feet downstream. Hold him in a bear hug while keeping his face close to yours. The idea is to provide a little extra buoyancy. Once you reach calmer water you can swim the patient to shore using a cross-chest carry or by grasping his hair or helmet.

Cardiopulmonary Resuscitation (CPR)

CPR is a proven life-saving technique that every whitewater paddler should know. Probably the most important thing about CPR is to start it as soon as possible, if the victim is not breathing, even if this means doing so in the water when you first make contact. Your first step will be to open the victim's airway by pulling up on his neck with one hand and pushing down on his forehead with the other. This tilts the head back and may cause the patient to start breathing again on his own. This you can often do in the water, but do not attempt compressions while in the water: get the patient to the nearest hard surface. This could be boulders in mid-stream, gear boxes on a loaded raft, or even the hull of a canoe. Do not spend precious moments hauling the victim up a river embankment if you can find a flat place at the water's edge. If you are alone, initiate one-person CPR until help arrives. (For details on CPR, see Appendix H.)

After starting CPR, keep doing the best possible job for at least an hour before attempting to transport the patient, since CPR is difficult to perform effectively while you are moving. In an emergency like this you will feel urgently that you want to get the

Fig. 9.1. For rescues in heavy water, use the bear hug (A) to add buoyancy and keep contact with the victim, then use the cross-chest carry (B) in calmer water to swim the victim to shore.

patient to an ambulance or hospital, but don't jeopardize the quality of your CPR in a rush to evacuate. If the patient has not responded within an hour and medical assistance is not nearby, it is unlikely he will survive anyway.

If CPR is successful and the patient begins to breathe on his own he must be handled as gently as possible. Rescuers must stand by to restart CPR at any time. Keep the patient warm, treat him for shock, and evacuate him. Under no circumstances allow a paddler to get back in his boat. If you are forced to spend the night in the field, the patient's condition should be continuously monitored, preferably by at least two people. Anyone who has had to have CPR administered should be admitted to a hospital for observation, no matter how well he might feel.

Paddlers should be aware of the Mammalian Diving Reflex (MDR). Sudden immersion in cold water (70°F or less) sometimes triggers a nervous response that slows the heart rate and concentrates circulation between the brain and heart. All body functions are slowed to a barely discernible point: the body is sustained on the oxygen present in the blood and tissues at the time of immersion. To an observer the victim may appear dead: he has very slow and shallow respiration; a weak, slow pulse; fixed, dilated pupils; and clammy, white skin. But MDR victims have been revived without brain damage after being submerged for as long as 38 minutes, so don't give up on rescue efforts or CPR just because four minutes (the old standard of "clinical death") have passed. If the victim is young, the water is cold (less than 70°F) and clean,

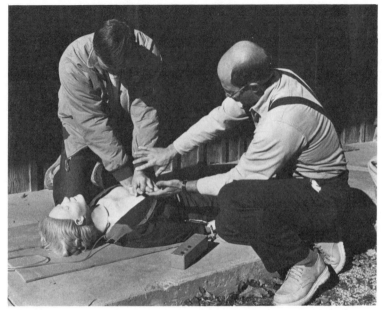

Fig. 9.2. Every paddler should take a course in cardiopulmonary resuscitation (CPR).

and the submersion time is less than one hour, MDR may have occurred and you should start CPR as soon as possible.

Hypothermia

Hypothermia is the cooling of the body to such a point that it can no longer maintain a constant temperature. Paddlers are mainly concerned with immersion hypothermia, which is caused by cold water flowing around the body, but wind-chill hypothermia can also be a factor.

The best way to deal with hypothermia is to prevent it. Wear the proper clothing and have the skill to paddle the water you've chosen. Know the warning signs of hypothermia (see Appendix I), and take action to keep matters from worsening as soon as you recognize them. Hypothermia affects judgment and coordination and is insidious in its effects. A good example is the paddler who swims once and gets cold, then starts missing his roll and swims again and again, becoming colder and more exhausted each time until finally he becomes hypothermic. Keep an eye on people after they swim. If it happens again right away and they are shivering, it is time to suggest a warm-up.

What is the best treatment for hypothermia on the river? In the early stages the patient can walk, and that's exactly what he should do. Walking will rewarm him and get him away from the probable cause of his hypothermia — the cold water. Consider walking out from the site, but don't let the patient do it alone: hypothermia is notorious for clouding a person's judgment, and people have got lost on simple trails because of it. At least one and preferably two people should accompany the patient.

When walking, do not ignore wind chill, rain, and other factors. There have been several fatalities when paddlers have tried to walk out in very poor conditions and never made it. If you are carrying basic survival gear (including matches) it may be better, especially if you can find a protected spot, to try to rewarm the patient on the spot while someone goes for help. Build a fire and if possible get the patient into dry clothes or a sleeping bag. If these aren't available, use extra clothing from other members of the group. Give the patient a warm, non-alcoholic drink ("snake-bite medicine" will only make the situation worse).

This treatment will not work in the second stage of hypothermia (body-core temperature: 90° to 95°F). The point at which hypothermia becomes critical is when the body cannot rewarm itself without outside help, even in a dry sleeping bag. One solution at this point is to use body heat from other members of the group in a "human sandwich." Skin-to-skin contact is necessary if this is to work. Another solution is to heat moist lifejackets in front of a fire and wrap them around the patient, exchanging them for freshly warmed ones as they cool.

The more advanced the hypothermia, the less effective field rewarming techniques will be. In the advanced or third stage (body-core temperature: 90°F or less) there is a danger of cardiac arrest if the patient is suddenly rewarmed (as by being put into a heated car or cabin), since this causes the stagnant, chemically unbalanced blood from the limbs to start recirculating again. This stage of hypothermia is on the whole easily recognized: the patient is unconscious. Field rewarming is impractical and the patient *must* be evacuated to a hospital. The limbs should be left uncovered but the torso kept warm. As with CPR, though, don't give up: people have been revived with a core temperature of 64°F and no heartbeat.

Shoulder Dislocations

Whitewater paddling can cause a variety of injuries: broken limbs and noses, sprains, cuts, and others. A common injury, particularly among kayakers, is a dislocated shoulder. A shoulder dislocates when the ball of the upper arm pops out of the socket of the shoulder. It is usually very painful and under normal circumstances should only be "reduced" (put back in place) by a doctor.

Shoulder dislocations are frequently caused by poor paddling technique. A typical case is when a paddler extends his arm away from his body, rotates it rearward on a brace, turns his head in the opposite direction, and receives a jarring blow. All paddlers should take the time to learn proper paddling technique and methods of avoiding shoulder dislocations.

First aid field treatment is to put a sling on the arm and then tie a swath bandage around the body to keep the arm immobilized. The patient should be evacuated, and although he can usually

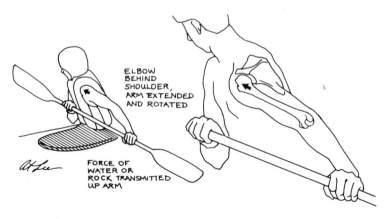

ELBOW
BEHIND
SHOULDER,
ARM EXTENDED
AND ROTATED

FORCE OF
WATER OR
ROCK TRANSMITTED
UP ARM

Fig. 9.3. This is how a shoulder dislocation happens.

Fig. 9.4. A dislocated shoulder is especially common among kayakers.

walk he should not be sent out alone: pain from a shoulder dislocation can induce shock.

Paddlers venturing into remote areas may want to discuss methods of reducing a shoulder dislocation with their doctor. This is especially true of those who have suffered a dislocation in the past, since they are more likely to have it happen again. The longer the shoulder remains "out" the harder it is to put it back in.

Evacuation Techniques

As the first aider, you are the one who must decide whether the injured party should be evacuated or not. This is sometimes not an easy decision, and you should only make it after considering all the alternatives. Will the patient's condition worsen with transportation? or is the injury so critical that an immediate evacuation is essential? If the party is small, help may have to come from an outside source. Some injuries require special equipment, like a backboard for back injuries or a traction splint for a fractured femur. In some cases, especially if the injury is not severe, the patient is embarrassed "to cause so much trouble" and will want to

Fig. 9.5. Deciding whether and how to evacuate a patient is not always easy.

continue the trip. You may have to insist that his evacuation is for the good of the trip as well as his own health.

When planning the evacuation, confer with the first aid and evacuation teams to determine the best evacuation route. Some routes may be automatically ruled out once you have carefully considered the nature of the injury. If the injury is not disabling, the patient can often walk out under his own power, and this is certainly easier than carrying him. But his condition could worsen, or he could get disoriented and lost or go into shock. As a precautionary measure, at least two people should be sent out with him, so that if the patient's condition does worsen one can stay with him and another can go for help. The first aider should stay with the patient all the way to the hospital. He is familiar with the injuries and vital signs and can monitor change more accurately. Perhaps more important, he and the patient will have developed a rapport that should not be interrupted.

In a very remote area, or if speed is essential and the surrounding terrain rough, or in a canyon, where overland evacuation is virtually impossible, the best means of evacuation may be the river (another alternative, helicopter evacuation, is discussed in Chapter 7). Most patients with serious injuries are not anxious to get back on the water, and the flexing of an inflated boat can be painful even to the best-splinted fracture, but this may be the only alternative.

If you choose river evacuation, keep the following considerations in mind:

- Never tie a patient into a raft or into a litter in a raft. In addition to the real danger of drowning if the raft flips, it is important to consider the patient's feelings of fear.
- Two canoes can be lashed together, catamaran-style, for increased stability.
- You can line the easier rapids and carry the patient around the more difficult ones.
- If you must cross above dangerous rapids or falls, rig a belay line or Telfer lower.

Lightly injured people can be carried for short distances with one- or two-man carries: the clothes drag (Fig. 9.6) is for emer-

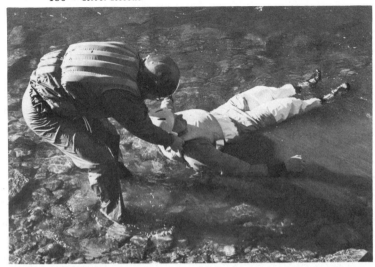

Fig. 9.6. Clothes drag.

Fig. 9.7. Fireman's carry.   *Fig. 9.8. Cross-shoulder carry.*

Fig. 9.9. Two-handed carry (piggyback).   *Fig. 9.10. Two-man carry.*

gency use only; the fireman's carry (Fig. 9.7) or the cross-shoulder carry (Fig. 9.8) are often effective for transporting people from the water to shore; the piggyback (Fig. 9.9) is really only practical with light patients; the two-man carry (Fig. 9.10) requires a wide path. One of the most comfortable ways to carry someone is to use the rope coil, a technique borrowed from our climbing friends. You must have a rope long enough (60 feet or more) to put into a climbing coil (Figs. 9.11-9.12). The coiled rope then goes around the rescuer's shoulders so that the patient can sit in it.

Injuries of a more serious nature require the use of a litter. Ready-made litters are ideal if available (some outfitters place them near potential trouble spots ahead of time). If they are not available, you can construct them from all kinds of materials — all it takes is a little imagination. Here are some examples:

- A rope litter can be improvised from two throwropes. "Weave" it with a series of slip knots (Fig. 9.14).

Figs. 9.11-9.12. Two views of the rope-coil carry.

Fig. 9.13. A canoe can be used as an evacuation litter by towing it sideways along a railroad track. Many rivers have railroad tracks alongside them.

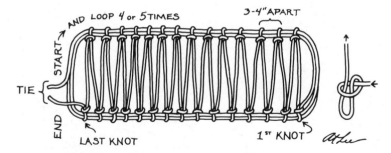

Fig. 9.14. Woven rope litter.

- Lifejackets can be rigged on saplings or paddles.
- A canoe with the flotation removed can be used as a litter. A canoe slid crossways along railroad tracks, using two rescuers as "dog teams," makes an excellent evacuation vehicle (Fig. 9.13).
- A kayak or C-1 can also be used as a makeshift litter, as long as the walls and seat are removed. This works best with higher-volume boats.
- Your imagination is the limit. River debris such as car hoods, lumber, and billboards have been used in emergencies.

Moving the Litter

Evacuations are hard work. The primary goals of the litter team are to keep the patient horizontal and to make the movement of the litter as smooth as possible.

Test any litter first with a healthy person. Designate one of the litter bearers captain. This person will coordinate lifting, lowering, and movement. You should by now have selected and scouted the evacuation route, but it is still a good idea to have a scout in front of the litter team to pick the exact route around boulders, trees, and other obstacles.

More than six litter bearers will get in each other's way: other team members should stand by to relieve the bearers from time to time. The safest way to switch bearers is to put the litter down, but this takes time. A more expedient method is to switch one person at a time "on the fly." Relief bearers walk alongside and take over

Fig. 9.15. Decked boats can also be used as evacuation litters, although the walls and seat must usually be removed. It's helpful to have someone stay with the boat to keep it from flipping over.

from the litter bearers when they start to tire. Only one bearer should switch at a time, and litter bearers should switch frequently, so that they don't get so tired they might drop the litter. Shoulder straps help distribute the load (Fig. 9.17).

If the litter team is faced with an obstacle the bearers cannot step over, the bearers should stay in one place and pass the litter to another team on the other side with a caterpillar pass (Fig. 9.16).

In rugged terrain the patient must be tied into the litter. Foot straps, groin straps, and underarm restraint straps will keep him secure. Pad the straps and check the patient occasionally to see that everything is comfortable. If the patient is conscious, don't tie his arms: he can fend off branches, scratch itches, and generally not feel so helpless. Be sure, though, to tell the patient to hold on to the straps rather than the side of the litter — remember the incident described at the beginning of this chapter.

When moving a litter up a steep bank, something you often have to do in a river evacuation, a talus belay is a useful technique (Fig. 9.18). Tie two of your longest ropes to the head of the litter,

Fig. 9.16. An evacuation in progress. In extremely rough terrain like this, litter movement is a series of caterpillar passes. Good organization and route scouting are important. (Photo by Ellyn Feinroth/Nantahala Outdoor Center)

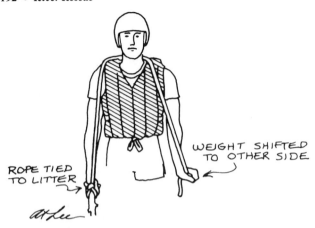

ROPE TIED
TO LITTER

WEIGHT SHIFTED
TO OTHER SIDE

Fig. 9.17. A shoulder strap makes litter carrying easier.

then send a belayer as far as possible up the bank to select a belay point like a tree or boulder. The belayer should keep the rope taut as the litter is moved up the hill, so that if it falls the rope will keep it from bobsledding back down the bank. When the litter gets to the first belayer, a second belayer should repeat the process from the next belay point up the bank. In this way, the litter is always secured and you lose little time. This technique can also be used for moving a litter downhill.

Your responsibility to your patient doesn't automatically end when the evacuation team reaches an ambulance or rescue squad. In remote areas (particularly outside the United States), you may find volunteer rescue squads or ambulance personnel whose hearts are in the right place but who lack up-to-date training. Do they look organized, well-supplied, and medically competent? Most squad members and ambulance drivers should be EMTs (Emergency Medical Technicians) and be trained and equipped for patient care. If the medical authorities do not appear competent, you must continue care of your patient until you find someone who can provide adequate care. This may require you to be direct and insistent about the manner of treatment and can be difficult if you are dealing with a legally constituted civil authority like a local search-and-rescue squad. If possible, the original first aider should accompany the patient to the hospital.

Fig. 9.18. The Talus belay alternates belay points to provide security against the litter slipping back down the bank. The litter should not be moved without one of the ropes on belay.

After any evacuation, take some time to discuss what happened. Mistakes can be identified while they're still fresh in everyone's mind. If the evacuation has been long and difficult and it's late, consider terminating the trip for the day.

Conclusion

Evacuations are no fun, but they are a very real part of whitewater rescue and every paddler should have some basic knowledge of them. Professional outfitters and whitewater clubs often practice evacuations at potential trouble spots and have rescue gear (pulleys, backboards, first aid supplies) cached there. Quick, smooth evacuations save lives.

*Training is everything. The peach was once a bitter almond; cauli-
flower is nothing but cabbage with a college education.*

MARK TWAIN, *PUDD'NHEAD WILSON*

· 10 ·
The Professionals

So far, this book has been primarily directed toward the
recreational whitewater user. We have described rescue systems based on equipment that a typical paddler might carry on a
river trip and have not really discussed the specialized equipment
and training that is the concern of the professional.

By "professional" we mean a person who makes his living on the
river: i.e., a guide, instructor, or boatman. These we will call river
professionals. By guiding for hire the river professional assumes
an added responsibility for the safety of his customers, who are
therefore entitled to rely on his expertise, training, and judgment. But there is another important type of professional: the
search-and-rescue squad member, the firefighter, or the park
ranger whose duty it is to rescue members of the public in trouble,
and this large category of "rescue professionals" includes both
amateurs and professionals, in the sense that some are paid to do
their rescue work full time and others, such as volunteer fire
departments and rescue squads, are unpaid part-timers. Repre-
senting a variety of agencies, these people try in general to prevent accidents and save lives, but their level of training and
expertise in whitewater rescue varies greatly.

In this chapter we will discuss some things both types of professional should consider about safety and rescue in their day-to-day
operations. However, neither type exists in a vacuum, and each

has areas of expertise and types of equipment that can benefit the other. Cooperation can be a problem: the rescue professional often ignores the wealth of experience of the river professional and regards him as an amateur, while the river professional too often regards the rescue professional as a threat to his operation. Especially on public lands, where river companies are regulated by state or federal agencies, cooperation and understanding are essential, so we urge each type of professional to read the other's section of this chapter to get an idea of what it is reasonable to expect.

The Rescue Professional

Professionals in almost any state may someday be faced with a whitewater rescue. About ten percent of the people who drown in the United States each year are rescue professionals who die in the line of duty. Preparation and training are vital.

Fig. 10.1. River and public-service professionals can work together. Here a National Park Service ranger gets instruction on setting up a Z-drag in an outfitter-sponsored rescue clinic.

Specialized training in river rescue is becoming available for rescue professionals, but all would be well advised to study the earlier chapters of this book. There are many ways to do a whitewater or swiftwater rescue, and most of the techniques described in this book can easily be adapted to specific circumstances. At the very least, the rescue professional should be aware of the power of moving water and its hidden hazards and know how to protect himself before trying to save others.

Preparation for the public-service rescue professional means not only having the right equipment and knowing rescue techniques but also studying the rivers in his agency's jurisdiction. As far as rescue is concerned, rivers fall into two broad categories: whitewater rivers that are regularly used for recreation and other rivers and streams that are not normally swift but are subject to floods.

Many of this country's whitewater rivers are in state and national parks; some are even in urban areas. These natural resources represent access to adventure for many outdoor enthusiasts. How public servants deal with these resources has a direct effect on the number of accidents that occur on them. Resource managers should know the safe capacities and limits of the rivers under their jurisdiction. They should know the access points, construct gauges to measure water levels, and establish "safe" levels for recreation. It may be necessary to issue regulations concerning craft size and type and required safety gear. Information and warning signs can help the uninformed. Managers should identify hazards and disseminate information about them.

Where can resource managers learn about the river? Personal involvement and research are the best way. A lot of answers (and questions) come from the users, both recreational paddlers and professional outfitters. Outfitters are usually more conservative in their outlook, because of liability considerations and the need to protect rental equipment. Some expert recreational paddlers, on the other hand, will tend to minimize difficulties and overestimate the skill level of the casual paddler.

But what about the rivers, streams, and creeks that are not normally whitewater but which become so in flood? Nearby inhabitants often have a wealth of information about these rivers:

Fig. 10.2. River hazards, like highway hazards, can be marked on busy rivers.

where they flood their banks, which low bridges go under first, how big the waves get at the narrows, and much more. Managers should look at the river at its normal level. Are there large piles of debris piled up in the trees or in certain bends or constrictions in the river? These could be strainers at higher water. Are the banks undercut? Are there man-made hazards like low-head dams or fallen bridges? What is the gradient of that little stream, and what would it be like with ten times the flow? How big is the watershed and what are the water-absorbing qualities of the soil? Runoff from areas upstream may cause radical and unpredictable rises after a rain. Keeping records or finding those that other agencies may be keeping will help managers to know many of these things, as will observing a river in all its moods throughout the year. In-depth knowledge will help prevent accidents.

In many areas, the river itself is the boundary of an agency's jurisdiction. This can cause problems if no one considers in advance *which* agency is responsible for the rescue. When an Air Florida jet crashed into the Potomac River in 1982 the result was a classic case of jurisdictional confusion. In some cases, especially in urban areas, it is better to have one unit trained and equipped for river rescue and to have all calls referred to that unit, regardless of jurisdictional boundaries.

There are many things government agencies can do to prevent river tragedies. The state of Ohio, for example, after several drownings among recreational paddlers and their would-be rescuers on rain-swollen rivers, initiated an aggressive policy of accident prevention. This has in some cases taken the form of warning signs near potential high-water trouble spots on normally calm rivers; in other cases, actual physical barriers have been erected to force paddlers to portage around certain hazards like low-head dams. Man-made debris and defunct dams have been removed and some dams restructured to reduce the hydraulics at their base. Canoe liveries have been asked to place safety decals in their boats, and television programs have broadcast information on river levels and hazards.

A key aspect of successful river rescues is training. For most park rangers and firefighters, river rescue is a secondary aspect of their job. Because they do not normally work with whitewater they often lack an appreciation of its dangers and can unknowingly risk their lives in an emergency. Even routine whitewater maneuvers such as ferries and eddy turns must be practiced to be effective. Potential rescuers must know how to read water and spot potential capsize points, and they must be familiar with their equipment. However, the most important single aspect of any training program is probably teaching the rescuer to protect himself. He must know what constitutes a hazardous situation, how to avoid it, and, if avoidance fails, how to deal with it.

Having the right personal equipment (see Chapter 2) is also important. A good Type III or V lifejacket should be considered mandatory for river rescues: firefighters should *not* wear "call out" coats and boots. Everyone should know the basics of swimming in whitewater: keep the feet downstream, float near the surface, and swim aggressively for shore. This should be practiced in a strictly controlled situation before it is really needed.

Rescue courses such as the ones offered by the Nantahala Outdoor Center and others listed at the end of this book offer a safe and effective way for the rescue professional to learn from the river professionals.

Some Rescue Considerations for Professionals We realize that not all public-service professionals will be able to take river-rescue courses, so we offer the following recommended sequence of operations when faced with a whitewater or swiftwater rescue.

- Talk the victim into self-rescue, if this is a safe option. Often a frightened or disoriented victim will overlook an obvious self-rescue alternative. At the very least, maintaining communication with the victim will help keep his hopes up while the rescue is being readied. Let him know what you intend to do and how.

- Use a shore-based rescue if possible: do not jeopardize your safety or that of other rescuers unnecessarily. In addition to the techniques described in Chapters 5, 6, and 7, a fire-department team might use an extension-ladder or fire-hose rescue. In the latter, a fire hose is inflated (see description below) to become a semi-rigid, floating arm that can be extended to a victim in a low-head-dam hydraulic or stranded on a rock.

- A boat-*assisted* rescue is more hazardous than remaining ashore, but it is safer than direct-contact methods of rescue. The boat is used for ferrying tag lines, people, and equipment through calmer stretches of water above or below the rapid.

- A boat-*based* rescue requires contact with the victim. It is much more dangerous than the above systems because it exposes the rescuer to the same hazards as the victim. Consider the Telfer lower and direct-contact methods using tag lines and the like.

- A last-resort method is a one-on-one, in-the-water rescue: a rescuer goes in with wetsuit, mask, fins, and perhaps even scuba gear. This poses great danger to the rescuer and can only be used in relatively calm water. In any case, the rescuer should be experienced in whitewater swimming and be capable of self-rescue.

Professional Equipment In addition to the equipment standard to any rescue organization, the following items can be very useful:

Fig. 10.3. A rescue course at the Nantahala Outdoor Center.

- A line gun capable of shooting a light haul line 150 to 400 feet. Crossbow versions will shoot about 200 feet of line.
- Motor-driven or electric winches light enough to be hand carried to remote river sites. Tackle blocks and steel cable will increase their capacity. If winches are not available a simple "come-along" will often suffice and is usually superior to a Z-drag.
- Long rope bags with 150 to 300 feet of rope.
- "Scoop rigs" to scoop a victim from the water with a helicopter hoist.
- Adapter for inflating fire hoses. Air-valve fittings are available for standard fire hoses which allow them to be inflated to about 100 pounds per square inch. The hose can then be extended for a shore-based rescue.
- Inflatable rafts. Unfortunately, most rescue organizations use hard boats (e.g., a jon boat with an outboard

motor), which are not suited for whitewater use. Rafts are much more stable and forgiving and as we have described earlier lend themselves to a variety of rescue techniques. Their separate air chambers make them virtually unsinkable. For rescue use a good size is 14 to 16 feet long and 6 to 7 feet wide. The raft should be able to carry six to eight people and can be propelled with paddles, oars, or an outboard motor. If your organization does not have its own, you might be able to borrow one from a nearby outfitter.

The boats used by most rescue organizations are designed for rivers and lakes and are not suited for whitewater or swiftwater use. Jet props, on the other hand, can be used in rivers so rocky they would ruin a normal propeller. A normal outboard can be converted to a jet prop, which propels the boat with a jet of water, with a kit. If you plan to use a prop boat, choose it with power in mind rather than speed.

A new type of inflatable called a spider boat or cataraft has recently been introduced. It consists of two pontoons joined together in a catamaran-type rig. A net hangs between the two pontoon braces. The spider boat is very stable and unswampable. More testing is needed, but this may turn out to be the ideal river-rescue craft, especially for low-head-dam rescue.

All boats should be extremely careful in the backwash of low-head dams. The water in the backwash is very aerated and neither normal nor jet props will work well in it. Several firefighters have drowned while attempting to motor up to people held in a low-head-dam hydraulic.

The River Professional

The successful whitewater outfitter tries to run the "best" possible trip for his clientele. He wants his customers to enjoy the river experience, to be reasonably comfortable, and not to get hurt. He wants these people to return and knows that the words "safe" and "best" are in this respect interchangeable.

Responsible outfitters value their equipment and maintain it, knowing that they must periodically replace worn-out gear and

Fig. 10.4. First aid training is an important part of working professionally on the river. Skills must be practiced if they are to be effective.

Fig. 10.5. A good way to practice rescue and evacuation is with an "accident scenario." These guides are practicing rescue from a pothole. This is the same pothole, at higher water, as the one shown in Fig. 1.3.

Fig. 10.6. Guides must know evacuation routes in advance. Walking the evacuation trails is part of pre-season training for these guides.

rafts as they become unserviceable. They use the appropriate size of raft for the water they run and do not overload the boats so that they become difficult to control. Unfortunately, there are outfitters who do not do business this way. The popularity of whitewater rafting has attracted the inevitable marginal outfitters, who either cannot or will not put money into new equipment and guide training. Some, struggling to survive in a seasonal business, try to stretch the useful life of their equipment and employ poorly trained summer help.

Ideally, all guides should be thoroughly familiar with the river they will be working on, particularly if the trips are short, single-day trips. In multi-day, expedition-style trips this is not always possible, but responsible outfitters will select staff who are experienced in reading water and who have proven boat-handling skills. These people should have *current* CPR and first aid training. Trip leaders should have advanced first aid skills or be qualified EMTs. The outfitter's management should set standards and policies to ensure safe trips. Guidelines for safe water levels, weather conditions, and the portages of dangerous rapids should be established in advance and not left to the discretion of the trip leader.

Other useful management policies address things like camp sanitation, speed limits and vehicle care, drug and alcohol abuse, and regulations on public lands and rivers. Prudent outfitters will watch the policies of other companies carefully.

One area many outfitters are lax in is that of rescue training — it doesn't earn them any money and often they don't pay their employees for training time. But rescue skills are like any other skill: they atrophy with disuse. When the time comes, it is important that rescue knowledge be fresh in the rescuer's mind. Tyroleans, Telfers, and Z-drags are not the sort of thing you do every day; they should be relearned and practiced at least once a year, preferably before the spring high-water season.

Training like this is ideally done on the river on which you will work. Study rapids that have given trouble in the past and discuss which rescue system would work best. Each major rapid should have a general plan of rescue, including the evacuation route, worked out in advance. Often it is only by simulating the accident that you discover you will need an extra-long rope or some other odd piece of gear. Some outfitters cache backboards, medical supplies, and other special gear at rapids that have a history of accidents.

A good training technique is the "accident scenario." The trainer should simulate an accident and give the trainees all the particulars: the nature of the injury, weather conditions, number of people available, and so on. The appointed rescue leader must effectively organize his people and care for the patient. He may also have to unpin a raft or boat and organize an evacuation. After the scenario is played out, the participants debrief as a group. The experience gained can be invaluable.

The Nantahala Outdoor Center has taken the "scenario" technique a step further by organizing a competitive "River Rescue Rodeo." Individual skills like rope throwing, rescuing swimmers, and recovering swamped boats are timed and scored. Teams compete in a judged and timed rescue scenario. This is a game in which all paddlers are winners.

Conclusion

The skills the rescue professional needs are much the same as those the river professional needs. The rescue professional may have more and better equipment and support, but this is wholly or partly offset by the greater experience and knowledge of the river professional. All too often, pride is a factor on both sides: the search-and-rescue squad resents the amateurs in lifejackets trying to tell them how to do their job; the whitewater outfitters are often reluctant to call for help because they prefer to handle their own problems. The only solution lies in increased awareness of the abilities and skills of each team. The search-and-rescue community must become more aware of the special problems and skills needed for whitewater and swiftwater rescue; river professionals must improve their own skills and actively work to promote good relationships with local public-service professionals. Only in this way can all of us achieve our aim of safe, effective rescue.

Fig. 10.7. A "Rescue Rodeo" contestant takes aim for a rope throw. Events like this build rescue skills. (Photo by Cindin Carroll/Nantahala Outdoor Center)

AFTERWORD
by
Charlie Walbridge
ACA Safety Chairman

As interest in river recreation has grown, so has the number of unprepared paddlers. As a result, agencies responsible for public safety have been called upon to make swiftwater rescues with increasing frequency. Until fairly recently these operations were perilous at best: the basic techniques used by whitewater experts were not widely known in the search-and-rescue community. Few rescue units took any specialized training, and most of the rescues were improvised on the spot. The courage and ingenuity of these people notwithstanding, many of these attempts proved fatal to victims and rescuers alike.

Concerned with the backlash from these and other incidents and worried about restrictive legislation, professional river outfitters and whitewater clubs began to work to spread knowledge of their skills to those who needed them. Assisted by boating-safety agencies in several states, a body of knowledge was developed specific to the needs of the people in the field. We are now at the point where in some areas rescue professionals are capable of handling most situations and are constantly developing useful ideas based on their own operational needs. Sadly, this is not true in all areas of the country.

This book is aimed primarily at the whitewater enthusiast. If you have bought it for use in discharging your professional responsibilities, it will give you a good background not only to the basics of river safety and rescue techniques, but also an insight into the ways that rivers can be run responsibly. For more information on specific operational techniques suited to your needs, we suggest you contact the following sources of information and training:

Ohio Department of Natural Resources
Division of Watercraft
Fountain Square
Columbus, Ohio 43224

Pioneers in swiftwater rescue training, with emphasis on dealing with low-head dams. Manuals and courses available.

Rescue 3
Box 1686
Sonora, California 95370

A professional training organization giving courses primarily on the West Coast.

Nantahala Outdoor Center
US 19W, Box 41
Bryson City, North Carolina 28713

Whitewater professionals. Courses in river rescue for the recreational paddler and the search-and-rescue professional who must do on-site rescues.

Appendices

Appendix A
International Scale of
River Difficulty

(If rapids on a river generally fit into one of the following classifications but the water temperature is below 50°F or the trip is an extended trip in a wilderness area, the river should be considered one class more difficult than normal.)

Class I Moving water with a few riffles and small waves. Few or no obstructions.

Class II Easy rapids with waves up to 3 feet high and wide, clear channels that are obvious without scouting. Some maneuvering required.

Class III Rapids with high, irregular waves often capable of swamping an open canoe. Narrow passages that often require complex maneuvering. May require scouting from shore.

Class IV Long, difficult rapids with constricted passages that often require precise maneuvering in very turbulent waters. Scouting from shore is often necessary, and conditions make rescue difficult. Generally not possible for open canoes. Boaters in covered canoes and kayaks should be able to execute an Eskimo roll.

Class V Extremely difficult, long, and very violent rapids with highly congested routes that must nearly always be scouted from shore. Rescue conditions are difficult and there is significant hazard to life in the event of a mishap. The ability to execute an Eskimo roll is essential for those in canoes and kayaks.

Class VI Difficulties of Class V carried to the extreme of navigability. Nearly impossible and very dangerous. For teams of experts only, after close study and with all precautions taken.

Appendix B
Universal River Signals

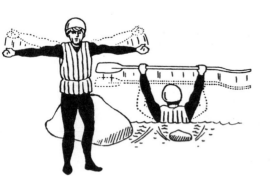 

Stop: *Potential hazard ahead. Wait for "all clear" signal before proceeding, or scout ahead.* Form a horizontal bar with your paddle or outstretched arms. Move this bar up and down to attract attention, using a pumping motion with the paddle or a flying motion with your arms. Those seeing the signal should pass it back to others in the party.

Help/Emergency: *Assist the signaller as soon as possible.* Give three long blasts on a police whistle while waving a paddle, helmet, or lifejacket over your head in a circular motion. If you do not have a whistle, use the visual signal alone. A whistle is best carried on a lanyard attached to the shoulder of a lifejacket.

All Clear: *Come ahead. (In the absence of other directions, proceed down the center.)* Form a vertical bar with your paddle or with one arm held high above your head. Turn the paddle blade flat for maximum visibility. To signal direction or a preferred course through a rapid or around an obstruction, lower the previously vertical "all clear" signal by 45 degrees toward the side of the river with the preferred route. Never point toward the obstacle you wish to avoid

Appendix C
The Force of the Water

The force of the water against an obstacle does not, as you might expect, increase in linear proportion to the velocity of the current. If a current of 3 cubic feet per second (cfs) exerts a force of 17 foot-pounds on your legs, you might reasonably think that a 6-cfs current would exert a force of 34 foot-pounds. *This is not so.* The force of the water increases in proportion to the *square* of the velocity of the current. Thus, if the current velocity doubles, the force of the water increases fourfold.

Current Velocity (CFS)	Average total force of the water (foot-pounds)		
	(ON LEGS)	(ON BODY)	(ON SWAMPED BOAT)
3	16.8	33.6	168
6	67.2	134.0	672
9	151.0	302.0	1512
12	269.0	538.0	2688

Appendix D
Useful Knots

Double Fisherman's Knot: Also used to tie together two pieces of rope, this knot is good when the ropes are substantially different in diameter.

Simple Prusik: Named for its inventor, Dr. Karl Prusik, this knot has many uses in river rescue. Its main virtue is that it will grip when under tension but stay loose when not.

Double Figure of Eight: Used to create a strong, easy-to-untie knot that makes a loop or bight in the rope.

Simple Bowline with Stopper: Easy to tie, this knot must be tied off or backed up in some way so that it will not loosen when not under tension. Feeding the free end back through the eye of the knot works well as a safety. This knot is not difficult to untie after use.

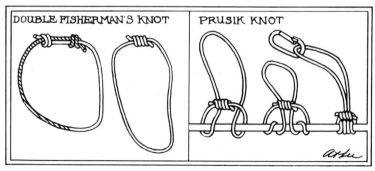

DOUBLE FISHERMAN'S KNOT

PRUSIK KNOT

DOUBLE FIGURE OF EIGHT KNOT/ROPE LOOP

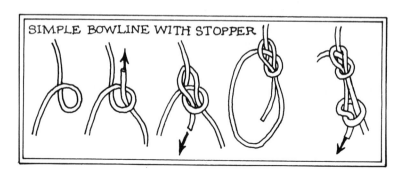

Appendix E
Cold Water Survival Chart

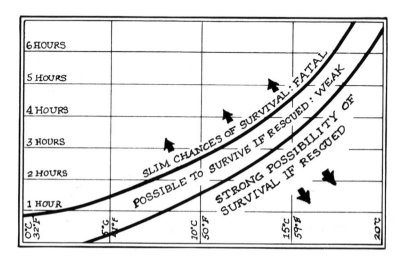

Appendix F
First Aid Kit Contents

1. "Ouch Pouch" for Day Trips

Ten 1″ adhesive bandages
24″ of half-inch adhesive tape
6 butterfly pads (3 large, 3 small)
8 gauze pads (four 4″ × 4″, four 2″ × 2″)
Roll of gauze (2″ wide)
10 aspirin or aspirin substitute
Sunscreen
Lip balm
Butane lighter
Wire saw
"Swiss Army" knife

2. Group Kit for Day Trips, Divided into Three Parts

Quick-access bag
20 aspirin or aspirin substitute
Ten 1″ adhesive bandages
Roll of adhesive tape
Small bottle Hydrogen Peroxide
4 large safety pins
Sunscreen

Tissue-injury Bag
Ten 1″ adhesive bandages
10 butterfly pads (5 large, 5 small)
20 gauze pads (ten 4″ × 4″, ten 2″ × 2″)
2 rolls of gauze strip (2″ wide)
Trauma dressing (gauze)
2 triangular dressings
2 tampons
2 elastic roller bandages (one 2″, one 3″)
Roll of adhesive tape
Chemical cold pack
Bottle of eyewash solution
Hand towel

Hardware bag
Scissors
Tweezers
Snake-bite kit (suction)
2 finger splints
Wire saw
Butane lighter
Small flashlight
Bee-sting kit (injectable epinephrine)
Emergency "space blanket"
Bottle of glucose
3 ammonia inhalants
Inflatable leg splint
Inflatable full arm splint
Nylon stretcher
First aid book

3. Expedition Recommendations

This kit would include all the contents of the group kit for day use with the number of individual items increased according to the length of the expedition and the number of participants. The additional items below require training in their use and should be varied according to the medical background of the members of the expedition.

Blood-pressure cuff
Stethoscope
Thermometers (oral and hypothermia)
Adult airway
Bulb syringe (for suction)
Suture kit
Intravenous kit
Drug kit
Tooth-fracture kit

Appendix G
Ground-to-Air
Signals for Survivors

These signals are intended for communication with a helicopter. They are advisory and a pilot is under no obligation to obey them. When these signals are used, it is important that the signaller be positioned beyond the path of the main rotor where he may be readily seen.

Clear to lift. Extend your arms horizontally, palms up.

Clear to start engine. Make a circular motion above your head with your right arm.

Move the helicopter back. Extend your arms forwards and "push" the helicopter away.

Hold on ground. Extend your arms horizontally, thumbs pointing down.

Move the helicopter forwards. Extend your arms forwards and wave the helicopter toward you.

Move to signaller's right. Extend your right arm horizontally and motion to your right with the palm of your left hand.

Cleared for take-off. Extend both arms above your head, thumbs up.

Move to signaller's left. Extend your left arm horizontally and motion to your left with the palm of your right hand.

Wave off: Do not land. Wave your arms from the side to over your head.

Release sling load. Touch your left forearm with your right hand, palm extended.

Land here: My back is into the wind. Extend your arms toward the landing area with the wind at your back.

Shut down. Cross your neck with your right hand, palm down.

The following signals are generally used for attracting the attention of the pilots of fixed-wing aircraft, though they may also attract a helicopter. Make the signs by placing rocks on the ground, spreading out clothing, scratching in the dirt, or stamping in the snow. In any case, make the signs large enough to be visible from high above.

∨ *Require assistance*

✕ *Require medical assistance*

N *No (or negative)*

Y *Yes (or affirmative)*

↑ *Proceeding in this direction*

Signals for pilots

Indicate wind direction. Circle.

No. Yaw back and forth.

Yes. Pitch up and down.

Appendix H
Cardiopulmonary
Resuscitation (CPR)

When a person's heart and lungs stop functioning because of shock, drowning, a heart attack, or other causes, it is possible to save that life by administering cardiopulmonary resuscitation, or CPR.

CPR provides artificial circulation and breathing for the victim. External cardiac compressions administered manually are alternated with mouth-to-mouth resuscitation in order to stimulate the natural functions of the heart and lungs.

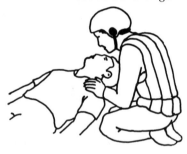

1. *Determine if the victim is unconscious.* Tap or gently shake the victim's shoulder. Shout "Are you OK?" If there is no response, shout "Help!" (Someone nearby may be able to assist.) Do the Airway step next.

2. *Airway step.* Place one hand on the victim's forehead and push firmly back. Place your other hand under the victim's neck, near the base of the skull, and lift gently.

Tip the head until the chin points straight up. This should open the airway. Place your ear near the victim's mouth and nose. LOOK at the chest for breathing movements, LISTEN for breaths, and FEEL for breathing against your cheek. If there is no breathing, do the Quick step next.

3. *Quick step.* Give four QUICK full breaths, one on top of the other. To do this, keep the victim's head tipped and pinch his nose. Open your mouth wide and take a deep breath. Make a good seal on the victim's mouth, then give the four breaths without waiting in between. Do the Check step next.

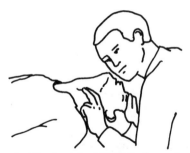

4. Check step. CHECK the pulse and breathing for at least five but no more than ten seconds. To do this, keep the victim's head tipped with your hand on his forehead. Place the fingertips of your other hand on his Adam's apple and slide your fingers into the groove at the side of the neck nearest you. If there is a pulse but no breathing, give one breath every five seconds. If there is no pulse *or* breathing, send someone for emergency assistance while you locate the proper hand position for chest compressions. Begin chest compressions.

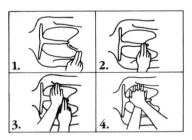

5. Hand position for chest compressions. (A) With your middle and index fingers find the lower edge of the victim's rib cage on the side nearest you. (B) Trace the edge of

the ribs up to the notch where the ribs meet the breastbone. (C) Place your middle finger *on* the notch, your index finger next to it. Put the heel of your other hand on the breastbone next to the fingers. (D) Put your first hand on top of the hand on the breastbone. Keep the fingers off the chest.

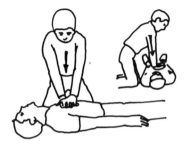

6. Chest compressions. *Push straight down,* without bending your elbows, always maintaining the proper hand position. Keep your knees a shoulder width apart. Your shoulders should be directly over the victim's breastbone. Keep your hands along the midline of his body. Bend from your hips, not your knees. Keep your fingers off the victim's chest. Push down 1½ to 2 inches. Count "1 and, 2 and, 3 and . . ." etc.

7. Push 15 — Breathe 2. Give 15 compressions at a rate of 80 per minute. Tip the victim's head so the chin points up and give two quick full breaths. Continue to repeat 15 compressions followed by two breaths. Check the pulse and breathing after the first minute and

every few minutes thereafter. NOTE: *Do not practice chest compressions on people — you might cause internal injuries.*

This information does not take the place of CPR training. Contact your local Red Cross chapter to find out how you can learn this life-saving procedure.

Appendix I
Symptoms of Hypothermia

Body Temperature: The hypothermia victim has a core (internal body) temperature that is lower than normal. The following table shows the successive stages of the condition:

Above 95°F The victim is conscious and alert and may have vigorous shivering.

90°-95°F The victim is conscious but has mild to moderate clouding of mental faculties. Shivering is present but diminished.

86°-90°F The victim has severe clouding of consciousness, may even be unconscious. Shivering is replaced by muscular rigidity.

Below 86°F The victim is conscious, with diminishing respirations.

Below 80°F The victim has barely detectable or non-detectable respirations.

Blood Pressure and Pulse: Blood pressure is lower than normal (frequently less than 100 mm mercury systolic). Pulse is generally slow and often irregular; it may be difficult to find it at all in the extremities because of blood-vessel constriction — measure the heart rate in the neck at the carotid artery or in the groin at the femoral artery.

General Appearance: The victim is pale in appearance and his skin is very cold to the touch. In fact, his skin and subcutaneous tissues are often at the temperature of the water he was immersed in. The victim's pupils begin to dilate at temperatures around 92°F and are fully dilated and poorly reactive to light at around 86°F.

Index

Accident prevention, 9, 199
"Accident scenario," 203, 205
Airbags, 50, 53-54
American Canoe Association (ACA)
    River Safety Task Force, 8, 32, 57
American Whitewater Affiliation
    (AWA) International Scale of
    River Difficulty, 22-23, **209**
Anchor line, 104-109
    passing a knot on, 148
    for Tyrolean rescue, 146-152
Anchor point, 106, 129, 137, 138
    for Tyrolean rescues, 146, 147-148
"Armstrong" method of boat re-
    covery, 119, **120**

"Baby bag," **46,** 103
Bear hug rescue, **178,** 179
Bechdel, Les, 2, 4-6, 135, 174-175
Belaying techniques
    body, 68, 72
    dynamic, 73, **74-75,** 80
    friction, 72, 110, 143
    static, **74-75,** 80
    talus belay, 191-193
Bernard, Rick, 3-4, 8
Berry, Karen, 160-163
Best of the River Safety Task Force

Newsletter, 32, 61, 157, 158
Bio-Bio River, 4, 174
Blackadar, Walt
    "Blackadar handle," **51,** 99
Boat-based rescue, 93, 168
    equipment retrieval, **94-99**
    Eskimo rescue, 93
    of swimmers, 99-104
    Telfer lower, **104-112**
Boat pins, 64, **114-116**
    extrication, 117
    prevention, 53, 54
    recovery, 119-134
    rescue techniques, 80-91
Boat recovery, 96-99
    See also Boat pins
Boats, decked, 37, 47, **49-51**
    as litters, 190
    recovery of, 96-97, 123
    rescue techniques, 93-94
    rescuing swimmers from, 99-100, 101
    self-rescue, 56-58, 64-65
    See also C-1s *and* Kayaks
Boats, rescue, 201-202
Boats, submerged, rescuing, 120-127
Boil line, **21,** 66, 67
Boulder sieves, 10, **19**

Bridge lowers, 86, **139-146**
  rigging for, 140-146
Broaching, 64, **114-115**
"Bulldozer" system of boat recovery, 51, **96,** 97
Bull Sluice, 12, 13
Butterfly knot, **87,** 127-128, 129

C-1s, 95, 116, 189
Canoes, open, 1, 47, 51, **53-55,** 188, 189
  entrapment in, 116
  recovery techniques, 97, 122-124
  rescue with, 100, 103, 106, 107, 109, 134
  self-rescue, 58, 61, 65
Capistrano flip, **58, 61,** 62
Carabiners, 35, 37, 38, **39-40,** 142, 143
  carabiner brake system, 143, **144,** 146
  carabiner chains, 107-108
  as pulleys, 127-128
Cardiopulmonary resuscitation, *see* CPR
Cataraft (spider boat), 202
Caterpillar pass, 191
Chattooga River, 3, 11, 61
  hazards, 12, 13, 15, 18, 19
  rescue on, 138, 176
Cheat River, 158
Chest harness, 140, **144**
Clearwater River, 32
Cleat, jam, **51,** 52, **53,** 98
Clothes drag, 185, **186**
Clothing, 42-45
Colorado River, 7, 11, 51
CPR, 118, 169, **178-180,** 204, **218-220**
Cradle rig, 121, **123-124**
Cross-chest carry, 178, **179**
Cross-shoulder carry, **186,** 187
Current strength, 125-126, **211**

Dams, low-head, 15, **19-21**
  rescue boats for, 202

rescue techniques, 66-67, 80-81
Death
  liability for, 173
  reactions to, 173-175
  what to do in event of, 171-172
Debris, 12, **19,** 66, 198, 199
"Deseret scale" for grading rapids, 22
D-rings, 48, 49, 60
Drowning, 3-4, 9, 19, 21, 61, 174-175
  "flush drowning," 22
Drysuits, 45

Eddy, 10, **14**
  maneuvering swimmer into, 70-71
Elevation loss, 11
Emergency kit, 41
EMTs (Emergency Medical Technicians), 192, 205
Entrapment, **16, 18, 19,** 114-116
  rescue techniques, 63-64, 80-88
  steps in rescue, 116-119
Equipment, 33
  maintaining, 202, 204
  preparing, 24
  professional rescue, 200-202
  repair, 26-27, 41
  retrieval, 94-99
  safety equipment, 26-27, **34-41, 45-47,** 199
Eskimo rescue, 93-94
Eskimo rolls, **56,** 57-58, 93
Eustis, Dick, 5
Evacuation
  of entrapment victims, 118-119
  helicopter, 158-160
  routes, 204, 205
  team, 169
  techniques, **183-194**
Extrication
  of entrapped victims, 117-119
  extrication team, 169
  haul systems, 126-135
  of pinned boats, 119-135

Eye bolt, 53

Ferries, 61
  rope, **81-83,** 86
Fiberglass boats, 1, 50
Fire hoses, 200
  adapters, 201
Fireman's carry, **186,** 187
First aid, 118
  kit, 41, **214**
  need for training, 177, 203, 204
  team, 169
  treatment for shoulder disloca-
    tion, 182-183
  See also CPR
Fixed-line rescue, 90, **91-92**
"Flip lines," 48, 58, **59**
Floating tag line, **80-83,** 117, 118
Flotation foam, 45
  blocks, 53-54
  walls, 49-50
Flotation ring, 83
Footbraces, 50
Foot entrapment, **16, 18, 63,** 80,
  84, 118

Gauley River, 12
Girth hitches, 147-149
Grab loops, 49, 50-51
"Grand Canyon" system of rating
  rapids, 22

Hammond, Drew, 5
Hare, Bruce, 4
Haul systems, 86, **126-135**
Helicopters, 119, 139
  evacuation by, 158-160
  rescue by, **153-158**
  "scoop rigs," 201
Helmets, 33, **37, 38**
High Country, 160
Holes, 10, **14-16,** 70
  self-rescue, 64-67
"Horizon line," 28, 29
Hydraulics, 10, 14, **15-16,** 19, 21
  rescue techniques for, **65-67,** 70,
  80, 99

Hypothermia, 22, 41, 42, 58, **180-
  182, 220**
  immersion, 42, 180-181
  rewarming techniques, 181-182
  wind-chill, 180

Injuries, 182-183

Jet props, 202

Karls, Bob, 157
Kayaks
  entrapment in, 114-116
  equipment for, 34, 36
  hazards of, 182-183
  as litter, 189
  paddle retrieval, 96
  rescues with, 100, 107
  self-rescue, 65
  See also Boats, decked
Kennedy, Payson, 4
Kern River, 156
Kerrigan, Dennis, 76
Knives, 35, 38, **39-41**
Knots, 39, 40
  bowline, 88, 89, **213**
  butterfly, **87,** 127-128, 129
  double figure-of-eight, 87, **131,
    154, 212**
  double fisherman's knot, 127,
    129, **131, 212**
  See also Loop knots; Prusiks

Landing zone, helicopter, 158, 159,
  160
Leader, trip, 26, **30,** 31
Leadership in rescue process, 164,
  166-167, 168, 205
Liability for death, 173
Lifejackets, 33, **34-37,** 38, 199
  modifying, 37, 41
  Type V, 36
  Type III, 34-36, 37
Line gun, 201
Litters, 187, 188, 189
  moving, 189, 191-193

switching bearers, 189, 191
Loop knots, 39, 87, **212, 213**

Mammalian Diving Reflex, 179
Mid-current lowers, 90, 91
Mirage, 52
Munter Hitch, 108, **111,** 112

Nantahala Falls, 23
Nantahala Outdoor Center, 2, 4, 5,
    23, 199, 201, 208
    "River Rescue Rodeo," 205-206
NOC knot trick, 87
Norton, John, 160-163

Ocoee River, 66, 135
    Powerhouse Rapid rescue, 139-
    140, 141, 160-163
Ohio Department of Natural Re-
    sources, Division of Water-
    craft, 83, 207
Organization, trip, 24-27, 30
Organization for rescue, 164-167
    teams, 168-171
Outboard boats, 202
Outfitting
    decked boats, 34, 49, 50, 51
    open canoes, 53-55
    rafts, 48-49
        double and triple rigging, 48-
        49
Overhand throwing technique, 78,
    79
Overhead rescue. *See* Vertical res-
    cue

Paddles
    proper technique for, 182
    retrieval, 95-96
    spare, 55
Painters
    boat recovery with, 97, 98, 100
    canoe, 54
Patient care, 176-183
    contact rescue, 177-178
    submerged victims, 179-180

Pendulum approach rescue, 89, **91**
Perrin, Dave, 4
Personal safety equipment, **33-41,**
    199
Piggyback (two-handed carry), 187
"Pitchpole" pin, 115
Portaging, 28, 29
Potholes, 13, **16**
Potomac River, 9
Preparation, trip, 24-32
Prusiks, **39,** 40, 131, **212**
    brake, 129-133
    traveling, 129-133
    Tyrolean rescue, 148
Public-service rescue professionals,
    195-196
    accident prevention, 199
    equipment, 200-202
    jurisdictional conflicts, 197-199
    training and rescue tips, 199-200
Pulleys
    traveling, 129-133
    Z-drag, 127-128

Rafts, 47, **48-49,** 201-202
    rescuing swimmers from, 103-
    104
    rigging for recovery, 123-125
    self-rescue, 58-61, 65-66
    use in Telfer lower, 104-105, 107,
    111
Ranger crawl, 149, 150
Ray, Slim, 2, 3-4
Repair kit, 26-27
Rescue, overhead or vertical, 86,
    139-152
    helicopter, **153-163**
Rescue professionals, **195-202,** 206,
    207
    equipment, 200-202
    public-service, 197-200
Rescue teams, 167, 168-169
    base camp, 170
    communication team, 169, 170
    evacuation team, 169, 170
    extrication team, 169, 170

first aid team, 169, 170
support team, 169, 170
Rescue 3, 208
Rescue training, 197, 199-202, 205
  "accident scenario," 203, 205-206
  courses, 199, 201
Rewarming techniques, 181-182
Rigging for pinned-boat recovery, 120-135
Righting watercraft, 58-59, 61
River characteristics
  elevation loss, 11
  importance of knowing, 22-23, 24-25, 197-199
  river-bed make-up, 11-12
  volume of flow, 11
River classifications, 22-23
  outfitting for Class I and II, 50
River evacuation, 185
River hazards, 10, 11, **12-22**
  big water, 22, 58
  cold water, 22
  identifying, 197-198, 199
River professionals, 1-2, 195, 202-206, 207
  management policies, 203, 204-205
Roll-over lines, 122-125, 128
Rolls, Eskimo, 56, 57-58
Rope
  attaching methods, 86-87
  ferries, **81-83,** 86
  how to hold, 74
  rescue by, 69-92
Rope bags, 201
Rope-coil carry, 187-188
Rope-ladder rescue system, **152,** 154, 155, 156
Rope litters, 187, **189**
Rope loops, 127, 129, **131**
Ropes, 39, **45-47**
  for bridge lowers, 140-145
  grab loops, 49, 50-51
  for haul systems, 127-134
  Tyrolean rescue, 146-152
  See also Throw bags

Rope tricks, **86-88,** 124

Safety, 7, 9, 26-27, 30, 57, 199, 204-205
Safety equipment, 33, **34-41, 45-47,** 199
"Scoop rigs," 201
Scouting, 28, 29
  evacuation route, 189
Sea-anchor haul system, 134
Self-lowering-rescuer system, 149, 151, 152
Self-rescue, **56-67**
Setnicka, Tim: *Wilderness Search and Rescue,* 160
Setting rope, 70, 71
Shock cord, 53
Shore-based lowering system, 151, 153
Shotover River, 157
Shoulder dislocations, 182-183
Shoulder straps for carrying litters, 191, 192
Sidearm throwing technique, 78
Sit harness, 140, **142-143,** 145, 163
Sliding seat lowering system, 149-151
Snag tags, 80, **83-86,** 118, 167
Spider boat (cataraft), 202
Standing wave, 10
Steve Thomas rope trick, 124
Strainers, 10, 16, **19,** 20
  self-rescue, 63, 70
Strong-swimmer rescues, **88-92,** 168
Stultz, Andrew, 4
Swamping, 51
  rescue measures for, 65-66, 112
Swimmers, rescuing, 69-75, 80, **99-112,** 168
Swimming, 58-62, 65-67, 199
  Capistrano flip, 58, 61, 62
  defensive, **61,** 62, 63-64

Tag-line rescues, 80-86
  floating tag line, **80-83,** 117, 118

snag tags, 80, **83-86,** 118, 167
stabilization tag line, 117, 118
Talus belay, 191-193
Telfer lower, 23, 86, **104-105,** 119
   lowering methods, 108-111
   rescue, 111-112
   setup, 106-108
"Ten Boy Scouts" method haul system, 126, 127
Thigh braces, 50, 54
   Thighstraps, 54, **55,** 161
Throw bags, **45-47,** 60, 69
   raft rescues with, 103
   throwing techniques, 75-79
Throwing rescue, 68-75
   with belaying techniques, 72, 73,
      74-75
   for multiple swimmers, 80
   techniques, 75-79
Throw ropes, 45-47, 69
"Throw sock," 47
Tow system, 51-53
   boat recovery with, 96, 97-99
Traveling pulleys, **129-133**
Trips, river
   choosing paddlers, 25, 30-31
   organization, 26-27
   preparation, 24-26, 33
   scouting, 28, 29, 32
   trip leader, 30-31
Two-handed carry (piggyback),
   187
Two-man carry, 187
Tyrolean rescues, 139, **146-152,**
   168

UH-1 series helicopter
   landing zone for, 158
Undercut rocks, 10, **16,** 17, 70
Underhand throwing technique,
   77-78, 79

Vector pull, 125-126, **127,** 128
Vertical pins, 115-116
Vertical rescue, 86, 139
   bridge lowers, 139-146
   helicopters, 153-160
   Tyrolean, 146-152
V-harness, 109, 110

Walbridge, Charlie, 32, 57, 61, 157
Watauga River, 113
Watercraft, 33, 47-55
   See also Boats, decked; Canoes,
      open; Kayaks; Rafts
Wetsuits, 44-45
"What If?" factor, 31-32, 166
Whistles, 35, 38, **41**
Whitewater
   history of river sport, 1, 7
   rating rapids, 22-23
   river characteristics, 11-12
Wilderness Search and Rescue (Setnicka), 160
Winches, 125, 127, 146, 155, 201
Wind direction, indicating, 159-160

Z-drag system, 39, 86, **127-133,**
   135, 136-138, 196
   and Tyrolean rescue, 147-148,
      149, 151, 153

ABOUT THE AMC

The Appalachian Mountain Club is a non-profit volunteer organization with over 30,000 members. Centered in the northeastern United States with headquarters in Boston, its membership is worldwide. The AMC was founded in 1876, making it the oldest organization of its kind in America. Its existence has been committed to conserving, developing, and managing dispersed outdoor recreational opportunities for the public in the Northeast. Its efforts in the past have endowed it with a significant public trust and its volunteers and staff today maintain that tradition.

Thirteen regional chapters from Maine to Washington, D.C., some sixty committees, and hundreds of volunteers supported by a dedicated professional staff join in administering the Club's wide-ranging programs. Besides volunteer organized and led expeditions, these include research, backcountry management, trail and shelter construction and maintenance, conservation, and outdoor education. The Club operates a unique system of eight alpine huts in the White Mountains, a base camp and public information center at Pinkham Notch, New Hampshire, a public service facility in the Catskill Mountains of New York, five full-service camps, four self-service camps, and nine campgrounds, all open to the public. Its Boston headquarters houses not only a public information center but also the largest mountaineering library and research facility in the U. S. The Club also conducts leadership workshops, mountain search and rescue, and a youth opportunity program for disadvantaged urban young people. The AMC publishes guidebooks, maps, and America's oldest mountaineering journal, *Appalachia*.

We invite you to join and share in the benefits of membership. Membership brings a subscription to the monthly bulletin *Appalachia*; discounts on publications and at the huts and camps managed by the Club; notices of trips and programs; and association with chapters and their meetings and activities. Most important, membership offers the opportunity to support and share in the major public service efforts of the Club.

Membership is open to the general public upon completion of an application form and payment of an initiation fee and annual dues. Information on membership as well as the names and addresses of the secretaries of local chapters may be obtained by writing to: The Appalachian Mountain Club, 5 Joy Street, Boston, Massachusetts 02108, or by calling 617-523-0636 during business hours.